If I Go Not Away

Mognu Masis

Published by Mognu Masis, 2026.

IF I GO NOT AWAY

First edition. January 10, 2026.

ISBN: 979-8233779275

Written by Mognu Masis.

Dedicated to the Wisdom of Eternal Awareness

Beyond Birth and Death

Perhaps the most terrifying duality for the human mind is the division between life and death, existence and non-existence.

We know, intellectually, that the body will die. But the mind recoils from this. We develop elaborate beliefs about afterlives to avoid confronting the possibility of annihilation. We distract ourselves with constant activity to avoid thinking about death.

But when you recognize your true nature as awareness itself, the fear of death loses its grip. The body-mind is born and will die, but awareness was never born and cannot die. It is the eternal background in which birth and death appear.

This is not a belief about immortality of the soul or reincarnation. It's a direct recognition that what you fundamentally are has no beginning or end. You are the timeless now in which the story of birth, life, and death unfolds.

If I Go Not Away

The Fundamental Truth of the Voice of Illumined

Introduction: The Great Forgetting

Right now, as you read these words, something is aware of them. Not just the words themselves, but the space around them, the sensations in your body, the quality of light, perhaps sounds in the background. Something is present to all of this. What is that something?

This question—simple, immediate, unavoidable—is the doorway to the most profound discovery available to human consciousness. Yet it's a discovery that isn't really a discovery at all, but a recognition of what has always been true, always been present, always been what you most fundamentally are.

Humanity finds itself in a peculiar predicament. We are divine consciousness having temporarily forgotten its true nature, playing out an elaborate drama of separation and materiality. This book explores the most profound paradox of human existence: how the infinite, eternal awareness that is the true nature of all reality came to appear as finite, mortal, and separate beings trapped in a world of matter and limitation.

The sages, mystics, and illumined beings of every spiritual tradition throughout history have attempted to communicate this fundamental truth. Yet their messages have been systematically misunderstood, literalized, and transformed into religious dogmas that often serve to obscure rather than reveal the very truth they were meant to illuminate. This misinterpretation represents what we might call "spiritual amnesia"—a collective forgetting of our true nature that manifests as religious literalism, scientific materialism, and the entire edifice of separation consciousness.

The purpose of this book is not to add another belief system to the countless philosophies and religions that already exist. Rather, it is to point directly to the living truth that precedes all concepts, beliefs, and intellectual frameworks. This truth cannot be fully captured in words, yet words can serve as fingers pointing toward the moon. The moon itself—eternal awareness, the ground of all being—must be directly realized through a shift in perception that transcends the intellect.

The Two Great Illusions

If this recognition is so available, why do most people miss it? Why does humanity remain caught in suffering, seeking, and the sense of separation? The answer lies in two great illusions that dominate human consciousness—illusions so pervasive that they're rarely questioned.

The first illusion is religious literalism—the tendency to mistake the finger pointing at the moon for the moon itself. When enlightened beings speak of God, Buddha-nature, Brahman, the Tao, or any other name for the ultimate reality, they are pointing to a direct experiential truth. But their followers, lacking this direct experience, convert these pointers into rigid belief systems, sacred texts, rituals, and institutions. The living truth becomes fossilized into doctrine. The pathless path becomes a highway with toll booths. The indescribable mystery gets reduced to theological propositions that must be believed on faith.

The second illusion is scientific materialism—the assumption that physical matter is the primary reality and that consciousness is merely an epiphenomenon of material processes. This worldview, while tremendously useful for technological manipulation of the physical world, commits the fundamental error of mistaking appearance for essence. It takes the dream seriously while denying the dreamer. It studies the contents of consciousness while ignoring consciousness itself.

Both illusions share a common root: the belief in separation. Religious literalism creates separation between the human and the divine, the sinner and the saint, the saved and the damned. Scientific materialism creates separation between the observer and the observed, the mind and the body, consciousness and matter. Both fail to recognize the singular, indivisible awareness that is the true nature of all apparent things.

The Three Fundamental Illusions

Within these two great illusions, three specific illusions create and maintain the sense of being a separate self:

1. The Illusion of Being a Separate Self

The first and most fundamental illusion is the sense that you are a separate, independent entity—a "me" that exists as a distinct thing in the world. This sense of being a separate self feels utterly obvious and undeniable. Of course you're a separate person! You have your own thoughts, your own body, your own life story.

But when you look carefully, can you actually find this separate self? Where is it located? Is it in your head? In your chest? Throughout your body? When you search for it directly in your experience, you can't find it as an actual entity. What you find instead is a collection of thoughts, sensations, memories, and perceptions—but no separate "self" that has these experiences. There is experiencing, but no separate experiencer. There is awareness of thoughts and sensations, but no separate entity that is aware.

The sense of being a separate self is actually a thought-construct, a story made of thought and memory, not your actual nature.

2. The Illusion of Time

The second illusion is time—the sense that you exist in a flow from past through present to future. You believe you were born in the past, you're living now in the present, and you'll continue into the future until you die. Your entire sense of identity is built on this temporal structure: your history, your story, your becoming.

But when you look at your direct experience, you never actually experience the past or future. You only ever experience now. The past exists only as present memory. The future exists only as present imagination or anticipation. The actual texture of your experience is always and only this present moment.

Even the sense of time passing is a thought arising now. The feeling of duration is a present sensation. When you look carefully, you can't find time in your direct experience—only the eternal present, the timeless now in which all experience arises and dissolves.

3. The Illusion of Being Contained in a Body

The third illusion is the sense that you—as awareness, as consciousness—are located inside your body, probably somewhere behind your eyes, looking out at a world that is separate from you.

But is this actually your experience? When you look carefully, where does your experience actually take place? You don't experience the world "out there" and yourself "in here." You experience a unified field of awareness in which all experience arises—sights, sounds, thoughts, sensations, the sense of the body, the sense of space.

The body appears in awareness. The world appears in awareness. Thoughts appear in awareness. But awareness itself has no location, no boundaries, no inside or outside. It's not contained in the body—the body appears in it.

This doesn't mean you are the body or that the body is an illusion. It means that awareness is not located in or limited to the body. Awareness is the open, spacious field in which the body and everything else appears. The body is a localized expression of awareness, not a container that awareness exists within.

You are not who you think you are. Your true nature is infinite, eternal, and identical with the source of all existence.

The Problem of Language

One of the primary challenges in discussing spiritual truth is the inherent limitation of language. Language evolved to help us navigate the world of separate objects, to distinguish this from that, to carve up the seamless wholeness of reality into manageable conceptual chunks. It is a tool designed for the domain of duality and separation.

But eternal awareness is non-dual. It is the indivisible wholeness that precedes and transcends all apparent divisions. Trying to capture non-dual truth in dualistic language is like trying to catch the ocean in a net. The net may catch some fish, but it cannot contain the water itself.

This is why the enlightened masters of every tradition have resorted to paradox, poetry, silence, and direct pointing. The Zen masters speak in koans designed to short-circuit the conceptual mind. The Christian mystics describe the "cloud of unknowing" and the "dark night of the soul." The Taoists remind us that "the Tao that can be spoken is not the eternal Tao." The Sufis use ecstatic poetry and whirling dance to bypass intellectual understanding.

Yet despite these warnings about the limits of language, subsequent generations of followers have attempted to systematize, codify, and literalize these teachings. The spontaneous utterances of realized beings get written down, compiled into scriptures, analyzed by theologians, and transformed into creeds and dogmas. The mysterious becomes the explained. The ineffable becomes the official doctrine.

This process of literalization serves certain social and political functions. It creates institutions with authority, hierarchies with power, and in-groups versus out-groups. But in the process, it obscures the very truth it claims to preserve and transmit.

The Spectrum of Understanding

We can think of spiritual understanding as existing on a spectrum from complete amnesia to full illumination:

Complete Spiritual Amnesia: At this level, there is total identification with the separate self and the material world. Consciousness is so fully absorbed in the dream of separation that it has completely forgotten its true nature. This is the default state for most of humanity most of the time.

Intellectual Understanding: Here, one has read or heard about spiritual truths and finds them intellectually interesting or compelling. One might believe in God, study philosophy, or practice meditation.

But this understanding remains conceptual—a collection of ideas about reality rather than a direct recognition of reality itself.

Glimpses and Insights: Occasionally, through meditation, prayer, psychedelic experiences, or spontaneous grace, the veil of separation temporarily lifts. One experiences moments of profound peace, unity, or recognition of a deeper reality. These glimpses are precious but temporary, and one inevitably returns to the ordinary state of consciousness.

Stable Realization: The shift from identification with the separate self to identification with eternal awareness becomes permanent. This is what traditions call enlightenment, liberation, or awakening. The person continues to function in the world, but from a radically different center of gravity. They know themselves to be the awareness in which all experience arises rather than a separate entity having experiences.

Full Illumination: This is the rarest attainment, where not only is the separate self seen through completely, but the entire appearance of the material world is recognized as the play of consciousness itself. Nothing is separate from the divine ground. Everything is seen as the one infinite awareness expressing itself in infinite forms.

The illumined beings whose teachings form the basis of the world's religions operated from the levels of stable realization or full illumination. But their followers, operating from levels of spiritual amnesia or intellectual understanding, inevitably misinterpreted their teachings through the lens of separation consciousness.

The Central Thesis

This book will demonstrate how the misinterpretation of enlightened teachings has created the false dichotomy between spirituality and science, between faith and reason, between the transcendent and the immanent. We will explore:

- How the non-dual teachings of Jesus were transformed into a religion of sin, guilt, and salvation through belief in historical events

- How the Buddha's direct pointing to the nature of mind became a complex philosophical system with its own dogmas and rituals
- How the radical monotheism of Islam, which denies all separation between creator and creation in its highest form, became identified with literal interpretations of scripture
- How the profound non-dualism of Advaita Vedanta has been both preserved and obscured within the Hindu tradition
- How the naturalistic mysticism of Taoism has been both corrupted by and compatible with modern scientific understanding
- How Jewish mysticism (Kabbalah) points to the same truths as other traditions while using very different language and symbols

We will show that what appears to be irreconcilable differences between religions, and between religion and science, are actually different perspectives on the same underlying reality viewed from different levels of understanding. When approached from the level of spiritual amnesia—whether in the form of religious literalism or scientific materialism—these perspectives seem contradictory. But when viewed from the level of awakened awareness, they are revealed as complementary descriptions of the one truth.

Most importantly, this book will not merely analyze these traditions intellectually. It will point directly to the possibility of shifting from identification with the separate self to recognition of your true nature as eternal awareness itself. For this is not a truth to be believed, but a reality to be directly realized.

Chapter One: The Original Message of Christianity

The Radical Non-Dualism of Jesus

To understand how Christianity became one of the world's great religions of separation, guilt, and belief in external salvation, we must first recover the original message of the historical Jesus—a message that was profoundly non-dualistic and centered on the direct realization of one's identity with the divine.

The central teaching of Jesus can be summarized in a few key statements that appear throughout the Gospels, particularly in the Gospel of John:

"I and the Father are one." (John 10:30)

"The Father is in me, and I in the Father." (John 10:38)

"On that day you will realize that I am in my Father, and you are in me, and I am in you." (John 14:20)

"The kingdom of God is within you." (Luke 17:21)

These statements are not claims to unique divinity that set Jesus apart from humanity. Rather, they are declarations of the fundamental truth of non-duality—that the apparent separation between the human and the divine is illusory. Jesus is not saying "I am God and you are not." He is saying "I have realized my identity with the divine, and you can realize the same truth."

This becomes even clearer when we consider Jesus's statement in John 14:12: "Very truly I tell you, whoever believes in me will do the works I have been doing, and they will do even greater things than these." How could this be possible if Jesus possessed some unique divine status that was fundamentally different from his followers? The implication is clear: Jesus had awakened to a truth about the nature of reality and identity that is available to all.

The Kingdom of Heaven

Jesus's primary teaching was about "the kingdom of heaven" or "the kingdom of God." This kingdom is not a location in the afterlife where good people go when they die. It is a state of consciousness available here and now to those who undergo a radical shift in perception.

"The kingdom of heaven is at hand," Jesus proclaimed. Not "will be available in the future" or "exists in another realm," but is at hand—immediately available, right here, right now.

When the Pharisees asked when the kingdom of God would come, Jesus replied: "The kingdom of God does not come with observation; nor will they say, 'See here!' or 'See there!' For indeed, the kingdom of God is within you." (Luke 17:20-21)

This is a clear statement that the kingdom is not an external place or future event, but an internal realization. It is the recognition of the divine presence that is already the ground of your being, obscured only by identification with the separate self and its thoughts, emotions, and desires.

The kingdom of heaven is what Jesus had realized. It is the state of consciousness in which one knows oneself to be one with the Father, in which the illusion of separation has dissolved, in which one sees God in all things and all things in God. It is, in the language of Eastern traditions, enlightenment or awakening.

Spiritual Rebirth

Jesus taught that entering the kingdom required being "born again" or "born from above." This has been literalized by many Christians to mean a one-time conversion experience in which one accepts certain beliefs about Jesus. But in the context of Jesus's overall teaching, being born again refers to a radical transformation of consciousness—a death to the old identity based on the separate self and a rebirth into the recognition of one's true nature.

"Very truly I tell you, no one can see the kingdom of God unless they are born again... Flesh gives birth to flesh, but the Spirit gives birth to spirit." (John 3:3-6)

This distinction between flesh and spirit is not a dualism between physical and non-physical, but between two levels of identity. The flesh represents identification with the body-mind organism, the separate ego that lives in time and dies. The spirit represents the eternal awareness that is your true nature, which was never born and can never die.

To be born of the spirit is to shift identification from the temporal self to the eternal Self, from the wave to the ocean, from the character in the dream to the awareness that is dreaming.

The Misinterpretation Begins

Even during Jesus's lifetime, his disciples struggled to understand his teaching. They kept interpreting his spiritual pointers through the lens of Jewish messianic expectations—the hope for a political-military leader who would restore the kingdom of Israel. When Jesus spoke of the kingdom of God, they heard a promise of worldly power and glory. When he spoke of his death and resurrection, they couldn't comprehend what he meant.

The real tragedy began after Jesus's death. The apostle Paul, who never met the historical Jesus, created a theological framework that shifted the focus from Jesus's teaching to Jesus's person. Instead of the message of non-dual awakening that Jesus taught, Paul emphasized belief in Jesus as the unique Son of God whose death served as a sacrifice for human sin.

This theological shift created several profound distortions:

1. Externalization of Salvation: Instead of salvation being the internal recognition of one's true nature (the kingdom within), it became something external—a transaction between God and humanity mediated by Jesus's sacrificial death.

2. Belief Over Realization: Instead of emphasizing the direct experiential realization that Jesus modeled, Christianity became centered on belief in certain propositions about Jesus.

3. Separation Theology: Instead of the non-dual message that "I and the Father are one" applies to all humanity, a radical separation was created between the divine Jesus and sinful humanity, with Jesus serving as the unique bridge.

4. **Future Orientation:** Instead of the kingdom being "at hand" and "within you," it became a future event—the second coming, the final judgment, the afterlife in heaven or hell.

These distortions were further amplified as Christianity became the official religion of the Roman Empire under Constantine. The diverse voices of early Christianity, including many that preserved more of the original non-dual teaching, were suppressed as "heresies." The Gnostic gospels, which often presented more mystical and non-dualistic interpretations, were excluded from the official canon.

The Christian Mystics: Preserving the Living Truth

Despite the theological distortions that came to dominate mainstream Christianity, the original non-dual message of Jesus was never entirely lost. Throughout Christian history, a stream of mystics and contemplatives have awakened to the same truth that Jesus taught and have attempted to communicate it within the framework of Christian language and theology.

Meister Eckhart, the 14th-century German mystic, spoke with remarkable directness about the identity of the soul with God:

"The eye through which I see God is the same eye through which God sees me; my eye and God's eye are one eye, one seeing, one knowing, one love."

"God and I, we are one in knowledge."

"The knower and the known are one. Simple people imagine that they should see God as if he stood there and they here. This is not so. God and I, we are one in knowledge."

These statements are virtually indistinguishable from the non-dual teachings of Advaita Vedanta or Zen Buddhism. Eckhart was, not surprisingly, charged with heresy. His teachings were too radical, too direct, too threatening to the institutional Church's claim to be the necessary mediator between God and humanity.

Saint Teresa of Avila described the highest stage of mystical union as one in which "the soul is one with God." Saint John of the Cross wrote of the complete transformation of the soul into the Beloved, such that "the two become one." The anonymous author of "The Cloud of Unknowing" pointed toward a knowing beyond all concepts and images, a direct encounter with the divine ground that transcends intellectual understanding.

These mystics understood that the spiritual journey is not about accumulating beliefs or performing rituals, but about a radical transformation of consciousness—a death to the false self and an awakening to one's true nature as one with God. They understood that the kingdom of heaven is not a place you go to after death, but a reality you awaken to here and now.

Contemplative Prayer and Meditation

The Christian contemplative tradition developed sophisticated practices for facilitating this awakening. Contemplative prayer, particularly in its highest forms, is remarkably similar to the meditation practices of Eastern traditions.

In the practice of centering prayer or the prayer of quiet, one lets go of all thoughts, images, and concepts—even concepts about God—and rests in silent, open awareness. This is not so different from the Zen practice of shikantaza (just sitting) or the Dzogchen practice of resting in the natural state.

The Desert Fathers and Mothers of early Christianity understood this. They withdrew to the desert not to accumulate theological knowledge, but to undertake the difficult work of seeing through the illusions of the separate self. Their sayings reveal a psychology remarkably similar to Buddhist teachings on the nature of mind:

"Abba Lot went to see Abba Joseph and said to him, 'Abba, as far as I can, I say my little office, I fast a little, I pray and meditate, I live in peace and as far as I can, I purify my thoughts. What else can I do?' Then the old man stood up and stretched his hands towards heaven. His fingers became like ten lamps of fire and he said to him, 'If you will, you can become all flame.'"

To become "all flame" is to be completely transformed, to have the separate self burned away in the fire of divine love, to become one with the source.

The Tragedy of Literalism

The tragedy is that this mystical stream has remained marginal in Christianity, while the literalistic, separationist interpretation became dominant. Most Christians have been taught to believe in Jesus rather than to follow his teaching and example toward the same realization he achieved.

They are taught that Jesus was the unique Son of God in a way that is fundamentally different from all other humans, rather than the exemplar who showed what all humans truly are when the veil of separation is lifted. They are taught to wait for salvation to come from outside themselves rather than to look within for the kingdom that is already present.

The result is that Christianity, which began as a path to awakening, became a religion of belief, guilt, and external salvation. The finger pointing at the moon became an idol to be worshiped. The map became mistaken for the territory.

This same pattern of literalization and misinterpretation appears in every major religious tradition. Let us turn now to Buddhism, where we will see how the Buddha's direct pointing to the nature of mind was transformed into complex philosophical systems and new forms of religious dogma.

Chapter Two: The Buddha's Original Teaching

Historical Buddha

Siddhartha Gautama, who became known as the Buddha (the Awakened One), was not trying to start a new religion. He was a radical empiricist who rejected the religious dogmas of his time and pointed to a direct investigation of the nature of consciousness and suffering.

When asked metaphysical questions—Does God exist? Is the universe eternal or not eternal? What happens after death?—the Buddha typically refused to answer. He compared such questions to a man struck by a poisoned arrow who refuses treatment until he knows who shot the arrow, what kind of wood it was made from, and what family the shooter belonged to. The point is not to accumulate metaphysical beliefs, but to end suffering here and now through direct insight into the nature of reality.

The Buddha's core teaching can be summarized in the Four Noble Truths:

1. Dukkha: There is suffering, dissatisfaction, and unsatisfactoriness inherent in conditioned existence.

2. Samudaya: This suffering arises from craving, attachment, and ignorance of our true nature.

3. Nirodha: It is possible to completely end this suffering.

4. Magga: There is a path that leads to the end of suffering.

This is not a theology or metaphysical system. It is a diagnosis and treatment plan. The Buddha saw himself as a physician treating the disease of suffering, not a prophet revealing divine truths to be believed on faith.

The Nature of Self

Central to the Buddha's teaching is the doctrine of anatta or anatman—no-self. This is perhaps the most misunderstood and misrepresented teaching in Buddhism.

Anatta does not mean that you don't exist or that consciousness is somehow an illusion. It means that what you take yourself to be—the separate, independent, continuous self—is a construction, a conceptual overlay on the immediate flow of experience. There is no substantial, unchanging self to be found when you look carefully at experience.

The Buddha invited people to investigate for themselves: Where is this self? If you look for it, you find only changing thoughts, sensations, perceptions, and consciousness itself. You never find a separate entity that possesses these experiences. The sense of being a separate self is a kind of optical illusion created by identification with the flow of experience.

This is remarkably similar to the pointing in Advaita Vedanta and other non-dual traditions. When you look for the self you think you are, you discover it has no independent existence. What remains when the illusion of the separate self is seen through? Pure awareness itself—the witness consciousness that is present in all experience but is not itself an object of experience.

Different Buddhist schools have described this differently. The Theravada tradition tends to emphasize the absence of self. The Mahayana tradition, particularly in texts like the Lankavatara Sutra, points more explicitly to the positive reality of Buddha-nature or the luminous nature of mind. The Vajrayana/Dzogchen tradition speaks directly of rigpa—the primordial awareness that is the true nature of mind.

Despite different emphases, all point to the same essential insight: what you think you are is not what you actually are. The separate self is a construction. Your true nature is the awareness itself.

Nirvana and Samsara

In the Buddha's teaching, nirvana is not a place you go after death if you've been good. It is the realization of the true nature of reality, available here and now. It is the cessation of suffering that comes from seeing through the illusion of the separate self and the grasping and aversion that arise from that illusion.

The Mahayana tradition developed this insight even further with the teaching that "nirvana and samsara are not two." Samsara—the world of suffering and illusion—and nirvana—the awakened state—are not different realities. They are the same reality seen from different perspectives. Samsara is what reality appears to be when seen through the lens of ignorance and separation. Nirvana is what the same reality is recognized to be when the veil of illusion is lifted.

This is why some Zen masters have said "ordinary mind is the way" or "nirvana is samsara, samsara is nirvana." They are not saying that suffering is good or that there is no difference between ignorance and enlightenment. They are pointing to the fact that awakening is not about going somewhere else or becoming something different. It is about recognizing what has always already been true.

The Eightfold Path

The Buddha's Eightfold Path is often presented as a moral code or set of religious practices. But in its essence, it is a comprehensive program for transforming consciousness and removing the obstacles to recognizing your true nature.

The path includes:

- **Right View** (understanding the nature of reality and causation)
- **Right Intention** (aligning one's motivations with truth)
- **Right Speech, Right Action, Right Livelihood** (ethical conduct that reduces harm and creates supportive conditions)
- **Right Effort, Right Mindfulness, Right Concentration** (meditation practices for training and stabilizing awareness)

This is not a set of commandments to be obeyed to please a deity or earn a reward. It is a pragmatic set of tools for the transformation of consciousness. Each element addresses a different aspect of the structure of suffering and separation.

Right View, for instance, is not about believing correct doctrines. It is about seeing clearly how suffering arises through ignorance, craving, and clinging. Right Mindfulness is not just a stress-reduction technique. It is the cultivation of moment-to-moment awareness that allows you to observe the arising and passing of experience without identification.

The entire path points toward the same destination: the direct realization of your true nature beyond the illusion of the separate self.

The Transformation into Religion

Despite the Buddha's emphasis on direct investigation and personal verification, Buddhism inevitably transformed into a religion with many of the same characteristics as other religious traditions.

The Deification of Buddha: In early Theravada Buddhism, the Buddha was understood as a human being who had awakened. But in Mahayana Buddhism, the Buddha became increasingly deified. The concept of the eternal Buddha-nature, the cosmic Buddha-bodies (trikaya), and celestial Buddhas and Bodhisattvas transformed Buddhism into something resembling the theistic religions the Buddha had originally moved away from.

This deification is not entirely without merit—it can be understood as pointing to the truth that the Buddha-nature or awakened awareness is the fundamental reality. But for most practitioners, it became another form of devotional religion with prayers, offerings, and reliance on external saviors rather than direct investigation of mind.

Philosophical Elaboration: The Buddha's simple, direct teaching about investigating the nature of suffering and the self became elaborated into increasingly complex philosophical systems. The Abhidharma texts analyzed consciousness and reality into minute categories. Madhyamaka philosophy developed sophisticated logical analyses of emptiness. Yogacara developed an idealist metaphysics about consciousness-only.

These philosophical developments have value and can be useful pointers. But they can also become obstacles—more content for the conceptual mind to grasp onto rather than tools for seeing through the mind's constructions.

Ritualization: Buddhist practice became increasingly ritualized, particularly in Mahayana and Vajrayana traditions. Elaborate ceremonies, prostrations, offerings, recitations of texts, and devotional practices became the primary forms of religious expression for most Buddhists.

Again, these practices can be skillful means when understood properly. But when they become ends in themselves, when practice becomes about accumulating merit or following tradition rather than awakening to true nature, they become another form of spiritual materialism.

Institutional Authority: Buddhism developed monastic institutions, hierarchies, lineages, and claims to authority based on tradition and institutional position. The spontaneous dharma of direct realization became codified into official teachings that must be transmitted through proper channels.

Literal Interpretation of Teachings on Rebirth: The Buddha's teachings on rebirth and karma have been literalized in ways that may distort their original intent. While the Buddha spoke of rebirth, it's worth questioning whether he meant literal reincarnation of a personal soul (which would contradict the teaching of no-self) or was pointing to the constant arising and passing away of the sense of self moment to moment.

When karma is understood as a cosmic justice system with literal rebirth in heaven or hell realms based on moral actions, it becomes another version of the reward-and-punishment theology found in theistic religions. This can create the same kinds of fear, guilt, and focus on external salvation that characterize religious literalism in other traditions.

Zen and the Return to Directness

Zen Buddhism represents, in many ways, a return to the directness and simplicity of the Buddha's original teaching. The Zen masters cut through philosophical elaboration and ritual complexity to point directly at mind:

"A special transmission outside the scriptures, not founded upon words and letters; by pointing directly to one's mind, one sees into one's true nature and attains Buddhahood."

When a student asked Joshu, "What is the meaning of Bodhidharma's coming from the West?" (a way of asking about the essential teaching of Buddhism), Joshu replied, "The oak tree in the garden." This is not a philosophical answer or a doctrinal statement. It is a direct pointing to the suchness of reality, before concepts divide it up.

The Zen tradition uses koans—paradoxical questions and stories that cannot be answered by the conceptual mind—to help students break through into direct realization. The koans are not puzzles to be solved intellectually. They are designed to exhaust the conceptual mind until it gives up, allowing direct insight to emerge.

Zen also emphasizes that enlightenment is not something gradual that you work toward over many lifetimes. It is available now, in this moment, because it is your true nature. This is the teaching of sudden enlightenment: the recognition that you have always already been what you are seeking.

Yet even Zen has not been immune to the process of literalization and institutionalization. Zen monasteries developed their own hierarchies and forms. The spontaneous dharma battles of the Tang dynasty masters became formalized koan study. The direct pointing became another tradition to be preserved.

The Universal Pattern

What we see in Buddhism is the same pattern we saw in Christianity:

1. An enlightened being directly realizes and points to the truth of non-dual awareness

2. A teaching emerges to help others toward the same realization

3. The teaching gets systematized, institutionalized, and ritualized

4. What was meant as a finger pointing at the moon becomes an object of worship in itself

5. The living truth becomes dead tradition

This pattern appears with remarkable consistency across all religious traditions. It reflects the inevitable tension between the truth that cannot be captured in forms and the human need for forms, structures, and traditions.

The question is: Can we recognize this pattern and learn from it? Can we receive the genuine pointing in these traditions while seeing through the literalistic misinterpretations? Can we distinguish the living water from the various containers that attempt to hold it?

Chapter Three: Islam and the Path of Unity

The Radical Monotheism of Islam

Islam's central teaching is captured in the Shahada: "La ilaha illallah"—There is no god but God. This is usually understood as a statement of monotheism, rejecting polytheism and idolatry. But at its deepest level, this declaration points to a radical non-dualism: there is nothing but God. All apparent separate existence is ultimately the one divine reality manifesting in infinite forms.

The Quran repeatedly emphasizes the absolute unity and transcendence of Allah while simultaneously declaring that Allah is "closer to you than your jugular vein" (Quran 50:16). This paradox—that the divine is both utterly transcendent and intimately immanent—can only be resolved in the direct realization that the apparent separation between creator and creation is illusory.

"To Allah belongs the East and the West, so wherever you turn, there is the Face of Allah." (Quran 2:115)

"We are closer to him than his jugular vein." (Quran 50:16)

"He is the First and the Last, the Manifest and the Hidden." (Quran 57:3)

These verses can be read as theological statements about an external deity. But they can also be read as pointers to the non-dual truth: there is only one reality, and that reality is both the source and substance of all that exists.

Sufism: The Mystical Heart of Islam

While mainstream Islam has often emphasized the transcendence of Allah and the separation between creator and creation, the Sufi mystics have preserved and developed the non-dual implications of Islamic teaching.

The great Sufi master Mansur al-Hallaj was executed for declaring "Ana al-Haqq"—"I am the Truth" or "I am God." His declaration was considered blasphemy by the religious authorities, who understood it as a claim that the individual ego was identical with God. But al-Hallaj was not speaking from the perspective of the separate self. He was declaring the truth of non-dual awareness: when the illusion of the separate self dissolves, what remains is the one divine reality.

Al-Hallaj explained: "I saw my Lord with the eye of my heart. I asked, 'Who are You?' He replied, 'You.'"

This is the same insight expressed by Jesus ("I and the Father are one"), by the Upanishads ("Thou art That"), and by the Zen masters ("Your original face before your parents were born"). It is the universal truth at the heart of all authentic spiritual realization.

Another great Sufi master, Ibn Arabi, developed the doctrine of wahdat al-wujud—the Unity of Being. This teaching holds that there is only one true existence, which is God, and all apparent separate existences are manifestations or self-disclosures of this one reality. The universe is the self-revelation of the divine.

Ibn Arabi wrote:

"There is no existence except His existence... The cosmos is His form and He is the spirit of the cosmos."

"My Beloved is three although He is One, even as the Persons are made One Person in the essence of the Trinity."

This last quote is particularly remarkable, as Ibn Arabi recognized the similarity between the Christian doctrine of the Trinity and the non-dual understanding of reality. When properly understood, the Trinity is not a mathematical impossibility (how can three equal one?) but a pointer to the mystery of unity in diversity—the one divine reality expressing itself in multiple forms while remaining one.

The Sufi Path

The Sufi path is designed to help the seeker move from separation consciousness to the realization of unity. This path typically involves:

Dhikr (remembrance): The constant repetition of divine names or phrases, particularly "La ilaha illallah," designed to replace ordinary thought patterns with awareness of the divine presence.

Fana (annihilation): The death of the ego-self, the dissolution of the separate identity. This is not a loss or diminishment but a liberation—the recognition that the separate self was always illusory.

Baqa (subsistence in God): After the annihilation of the false self, one continues to live and function in the world, but from the perspective of unity rather than separation. One subsists in God, recognizing that all actions are ultimately the action of the one divine reality.

This progression from separation to dissolution to abiding in unity is remarkably similar to the stages described in other mystical traditions. It is the universal path of awakening, expressed in Islamic language and symbolism.

Rumi, perhaps the most beloved of all Sufi poets, expressed this journey in terms of love and longing:

"The lover's cause is separate from all other causes. Love is the astrolabe of God's mysteries."

"I have put duality away, I have seen that the two worlds are one; One I seek, One I know, One I see, One I call."

"You are not a drop in the ocean. You are the entire ocean in a drop."

For Rumi, the spiritual journey is not about the human soul seeking union with a distant God. It is about recognizing that the apparent separation between lover and Beloved is illusory—they have always been one.

The Literalization of Islam

Despite the profound non-dual insights of the Sufis, mainstream Islam has generally moved in the direction of literalism and legalism, much like Christianity and Buddhism.

The Quran, which contains both legal prescriptions for 7th-century Arabia and profound mystical poetry, has been primarily interpreted through a literalist lens. The rich symbolic and metaphorical language has been flattened into propositions to be believed and rules to be followed.

The emphasis has been placed on external conformity to religious law (Sharia) rather than internal transformation. Prayer becomes a ritualistic obligation performed five times a day rather than the constant remembrance of the divine presence. The pilgrimage to Mecca becomes a physical journey rather than the inward journey to the heart where God is "closer than your jugular vein."

Most tragically, the doctrine of tawhid (divine unity) has been interpreted primarily as monotheism—the belief in one God rather than many gods—rather than as the non-dual truth that there is nothing other than God. This has led to a rigid separation between the sacred and the profane, the religious and the secular, the believer and the unbeliever.

The result is a religion of submission to external authority and law, rather than the direct realization of unity that the Sufis point to. The finger has once again been mistaken for the moon.

Chapter Four: Hinduism and the Teaching of Non-Duality

The Upanishads

The Hindu tradition is vast and diverse, encompassing everything from polytheistic devotional religion to the purest non-dualism. But at the heart of Hindu philosophy lie the Upanishads, texts that contain some of the most direct and explicit teachings on non-dual awareness ever written.

The central teaching of the Upanishads is expressed in four great statements (mahavakyas):

"Tat tvam asi" (Thou art That) - You are the ultimate reality

"Aham Brahmasmi" (I am Brahman) - I am the absolute

"Prajnanam Brahma" (Consciousness is Brahman) - Pure awareness is the ultimate reality

"Ayam Atma Brahma" (This Self is Brahman) - The true Self is the absolute

These are not philosophical speculations or religious beliefs. They are declarations of direct realization. The realized sages who spoke these words were not theorizing about the nature of reality—they were reporting what they had directly discovered when they looked into the nature of their own consciousness.

The Upanishads make clear that Brahman (the ultimate reality, the ground of all being) and Atman (the true Self, the deepest essence of consciousness) are identical. What you are seeking is what you are. The goal of the spiritual search is not in some distant heaven or future time—it is the recognition of your own true nature, here and now.

The Chandogya Upanishad tells the story of a student named Svetaketu who returns home after years of studying the Vedas. His father asks him, "Did you ask for that instruction by which we hear what cannot be heard, by which we perceive what cannot be perceived, by which we know what cannot be known?"

The father then teaches Svetaketu about the subtle essence that pervades all of reality, using the example of salt dissolved in water—invisible but present everywhere. After each teaching, the father concludes: "Tat tvam asi, Svetaketu"—"You are That, Svetaketu."

This teaching method is designed to shift identification from the body-mind to the awareness that witnesses the body-mind. You are not the changing forms that appear in consciousness. You are the consciousness itself, which is identical with the ultimate reality.

Advaita Vedanta

The non-dual teaching of the Upanishads was systematized and clarified in the tradition of Advaita Vedanta, particularly in the teachings of Adi Shankaracharya (8th century CE).

Shankara's teaching can be summarized in three key points:

1. **Brahman satyam** - Brahman (the absolute) alone is real
2. **Jagan mithya** - The world (of separate forms) is illusory
3. **Jivo brahmaiva naparah** - The individual soul is nothing other than Brahman

This requires careful understanding. Shankara is not saying that the world doesn't exist in any sense. He is saying that the world as we ordinarily perceive it—as a collection of separate, independent objects existing in their own right—is a misperception. The apparent multiplicity and separation is superimposed on the singular, non-dual reality of Brahman.

His favorite analogy is the rope mistaken for a snake. In dim light, you might see a rope on the ground and mistake it for a snake. You experience fear and react accordingly. But the snake has no independent existence—it is the rope misperceived. When you look more carefully and recognize it as a rope, the snake "disappears," though nothing has actually changed except your perception.

Similarly, the world of separate forms is Brahman misperceived through the lens of ignorance (avidya). When you recognize the truth, you see that there has only ever been Brahman, appearing as the many while remaining one.

Maya: The Creative Power of Consciousness

The concept of maya is often misunderstood as meaning that the world is simply an illusion, a kind of hallucination with no reality at all. But maya has a more subtle meaning.

Maya is the creative power of Brahman, the capacity of the one consciousness to appear as the many. It is both the power of veiling (which creates the appearance of separation and multiplicity) and the power of projection (which manifests the entire universe of forms).

Maya is not separate from Brahman—it is Brahman's own power. Just as the sun has the power to create light and heat, Brahman has the power to create the appearance of the manifest universe. The universe is real in the sense that it is Brahman's self-expression. It is illusory only in the sense that it appears to be multiple and separate when it is actually one and unified.

The Three States and the Fourth

Advaita Vedanta uses the analysis of waking, dreaming, and deep sleep to point to the nature of consciousness.

In the waking state, you experience the external world of objects and other people. This seems completely real and solid.

In the dream state, you experience an entire world that seems equally real while you're in it. You have a dream body, you encounter dream people and objects, you have dream adventures. Yet when you wake up, you recognize that the entire dream world was a projection of your own consciousness. The dream characters had no independent existence—they were all you.

In deep sleep, there are no experiences, no world, no separate self. Yet you don't cease to exist. When you wake up, you say "I slept peacefully" or "I didn't sleep well." You have some kind of awareness even in deep sleep, though it is not the awareness of any particular object.

Advaita asks: What is constant through all three states? The waking world comes and goes. The dream world comes and goes. Deep sleep comes and goes. But awareness itself is constant. There is a witnessing consciousness that is present in all three states without being identical to any of them.

This witnessing consciousness is called turiya—the fourth. It is not actually a fourth state in addition to the three, but rather the background awareness that is present in and through all states. It is your true nature.

The realization is: Just as the dream world is recognized upon waking to be a projection of your own consciousness, so too the waking world is ultimately a projection of the one consciousness that is your true nature. The apparent solidity and separateness of the waking world is no more ultimately real than the dream world.

This doesn't mean you should stop functioning in the world or treat it as if it doesn't matter. It means you recognize what it actually is: the play of consciousness, the dance of Brahman with itself.

The Path of Self-Inquiry

The primary method of Advaita Vedanta is self-inquiry (atma-vichara), most clearly articulated in modern times by Sri Ramana Maharshi.

The practice is simple: Ask yourself "Who am I?" and investigate the sense of being a separate self.

Are you your body? No—you are aware of your body. You can observe your body as an object of awareness, so you cannot be identical with it.

Are you your thoughts? No—you are aware of your thoughts. Thoughts come and go, but the awareness that witnesses them remains constant.

Are you your emotions? No—you are aware of your emotions. They arise and pass, but awareness continues.

Are you your memories? Your personality? Your roles and identities? No, no, no—all of these are objects appearing in awareness. None of them can be what you truly are, because you are aware of them.

So what are you? When you strip away everything you are aware of, what remains? Pure awareness itself—consciousness without an object, the witness that cannot be witnessed because it is the witnessing itself.

This is not a philosophical conclusion to be reached by logic. It is a direct recognition that becomes obvious when you actually look. You are not in consciousness as a separate entity having experiences. You are consciousness itself, in which all experiences arise and pass.

Ramana Maharshi taught that this inquiry should be pursued until the separate self is completely seen through and the natural state of pure awareness is stably recognized. He called this sahaja samadhi—the natural state. It is not a special altered state that comes and goes, but the recognition of what has always been true.

The Misinterpretation in Hinduism

Despite the clarity and directness of the Advaitic teaching, it has remained a minority view within the broader Hindu tradition. Most Hindus practice devotional religion (bhakti) focused on particular deities—Shiva, Vishnu, Krishna, Kali, and countless others.

This devotional practice is not wrong or necessarily contradictory to non-dualism. At a certain level, devotion to a divine form can be a skillful means for purifying the heart and directing attention toward the divine. The problem arises when the form is mistaken for the formless, when the deity is understood as an external being separate from oneself rather than as a representation of one's own true nature.

The Bhagavad Gita, Hinduism's most popular scripture, contains both devotional and non-dual teachings. Krishna tells Arjuna both "I am the source and dissolution of the whole universe" (9.18) and "Those who worship Me with devotion, they are in Me and I am in them"

(9.29). These can be understood as contradictory—is God a separate being to be worshiped or the ultimate reality identical with one's true self? Or they can be understood as teachings appropriate for different levels of understanding.

But for most Hindus, the non-dual teaching is too radical, too threatening to the sense of being a separate person with a separate life. It's more comfortable to worship an external God who might grant blessings and protection than to recognize that the worship and the worshiped are ultimately one.

The result is that Hinduism, like other traditions, has become largely a religion of ritual, belief, and external practice rather than direct realization. The profound non-dual wisdom of the Upanishads and Advaita Vedanta remains available but is often buried under layers of mythology, ritual, and social custom.

The Universal Pattern Continues

Once again we see the same pattern: Direct realization gets expressed in teachings. The teachings get systematized. The system becomes institutionalized. The institution becomes focused on preserving tradition rather than facilitating realization. The living truth becomes dead religion.

But in every tradition, the living truth has never been entirely lost. There have always been those who have awakened to the same truth, whether they called it union with God, enlightenment, fana, or moksha. The names differ, but the reality they point to is one.

Chapter Five: Taoism and the Nameless Way

The Tao that cannot be Spoken

The opening lines of the Tao Te Ching, the foundational text of Taoism, immediately point to the limitation of concepts and language:

"The Tao that can be told is not the eternal Tao.

The name that can be named is not the eternal name.

The nameless is the beginning of heaven and earth.

The named is the mother of ten thousand things."

This is as direct a statement as exists anywhere about the relationship between ultimate reality and the conceptual mind. The Tao—which we might translate as the Way, the Source, the Ground of Being—cannot be captured in concepts or language. Any attempt to define it, name it, or describe it necessarily falls short of the reality itself.

Yet there is a paradox here: the Tao Te Ching consists of 81 chapters attempting to point toward this that cannot be spoken of. Lao Tzu, the legendary author, uses language to point beyond language, concepts to point beyond concepts. This is the same strategy employed by mystics in all traditions.

What is the Tao? It is the nameless, formless, infinite source and ground of all existence. It is emptiness pregnant with all possibilities. It is the natural way that underlies and permeates all things. It is simultaneously nothing and everything.

Naturalness and Non-Action

A central teaching of Taoism is wu wei, often translated as non-action or effortless action. This is not passivity or laziness. It is the natural, spontaneous action that arises when one is aligned with the Tao rather than acting from the agenda of the separate self.

Watch water flowing downhill. It doesn't strain or effort. It simply follows its nature, taking the path of least resistance, yet nothing is more powerful. It wears away solid rock not through force but through persistent alignment with natural law.

Watch a tree growing. It doesn't try to grow. Growing is simply its nature when conditions are right. It doesn't attempt to be something other than what it is.

Wu wei is the action that arises from this kind of alignment with one's true nature, with the Tao itself. It is the dissolution of the separate will trying to impose itself on reality, and the emergence of spontaneous action from the ground of being.

This is remarkably similar to what other traditions describe as the action of the enlightened being. In Advaita Vedanta, actions continue after enlightenment, but they are understood to be the action of Brahman rather than the action of a separate individual. In Zen, this is described as "chopping wood, carrying water"—ordinary actions continue, but they arise from no-mind rather than from the separate self.

Emptiness and Fullness

The Tao Te Ching repeatedly uses the metaphor of emptiness to point toward the nature of the Tao:

"Thirty spokes share the wheel's hub;
It is the center hole that makes it useful.
Shape clay into a vessel;
It is the space within that makes it useful.
Cut doors and windows for a room;
It is the holes which make it useful.
Therefore profit comes from what is there;
Usefulness from what is not there."

The emptiness at the center is what makes things functional. Similarly, the emptiness that is the Tao is what allows all forms to arise.

This emptiness is not mere absence or void. It is pregnant with infinite potential, the formless ground from which all forms emerge.

This is the same insight as the Buddhist teaching of sunyata (emptiness) and the Advaitic teaching of Brahman as nirguna (without qualities). Ultimate reality is empty of all particular characteristics, and precisely because it is empty of all limitations, it can manifest as all things.

"The Tao is like a well:
used but never used up.
It is like the eternal void:
filled with infinite possibilities."

The Return to the Source

Taoism emphasizes that the spiritual path is not about going somewhere new or becoming something different. It is about returning to your original nature, to the source from which you never actually departed.

"Returning is the motion of the Tao.
Yielding is the way of the Tao.
All things are born of being.
Being is born of non-being."

You emerged from the Tao like a wave arising from the ocean. The wave has never actually been separate from the ocean—it is the ocean itself in wave form. The spiritual journey is the wave recognizing its true nature as ocean.

This teaching of return to the origin appears in many traditions. In Christianity, it's the return of the prodigal son to the father's house. In Buddhism, it's the recognition that samsara and nirvana are not two. In Advaita, it's the recognition that you have always been Brahman. In Sufism, it's the lover discovering they have always been one with the Beloved.

The Tao Te Ching expresses this beautifully:

"In the pursuit of learning, every day something is acquired.

In the pursuit of Tao, every day something is dropped.
Less and less is done
Until non-action is achieved.
When nothing is done, nothing is left undone."

The spiritual path, from this perspective, is not about accumulation but about subtraction. Not about becoming something you're not, but about letting go of what you never were. Not about attaining enlightenment, but about releasing the illusions that obscure your true nature.

Taoism and Science

Of all the classical spiritual traditions, Taoism may be the most compatible with a scientific worldview. It makes no claims about supernatural beings, miraculous events, or violations of natural law. Instead, it points to the fundamental nature of reality itself, which can be investigated directly.

The Taoist understanding of yin and yang—complementary opposites that arise together and transform into each other—anticipates modern understanding of polarity, feedback loops, and dynamic systems. The Taoist emphasis on natural processes and patterns resonates with ecology and systems thinking.

Modern physics, particularly quantum mechanics, has revealed a reality that seems more Taoist than Newtonian. The fundamental uncertainty and interconnectedness of quantum phenomena, the observer-dependent nature of measurement, the wave-particle duality—all of these echo Taoist insights about the fluid, interdependent nature of reality.

This doesn't mean ancient Taoists anticipated modern physics. It means that both are investigating the same reality, and both have discovered that reality is far stranger, more subtle, and more interconnected than the common-sense view of solid, separate objects in an absolute space and time.

The Literalization of Taoism

Like other traditions, philosophical Taoism became corrupted and literalized as it developed into religious Taoism. The nameless Tao became personified as deities. The natural practices for aligning with the Tao became magical techniques for gaining immortality or supernatural powers. The simple, direct pointers of Lao Tzu and Chuang Tzu became overlaid with complex rituals, alchemical practices, and popular superstitions.

This is the same pattern we've seen repeatedly: the living truth gets fossilized into religious forms. But the original teaching remains available to those who look for it.

Chapter Six: Jewish Mysticism and the Hidden Unity

The Esoteric Teaching in Judaism

Judaism is typically understood as a strictly monotheistic religion centered on the transcendent God (YHVH) who created the universe, chose the Jewish people, gave them the Torah, and demands obedience to commandments. From this perspective, there is an absolute separation between the creator and creation, between God and humanity.

But there has always been an esoteric stream within Judaism—the mystical tradition known as Kabbalah—that points to a more non-dual understanding. While using very different language and symbols than other mystical traditions, Kabbalah contains the same essential insights about the unity of all existence in the divine ground.

The Hebrew name for God, YHVH (often rendered as Yahweh or Jehovah), is derived from the verb "to be." God's response to Moses when asked his name is "Ehyeh Asher Ehyeh"—"I Am That I Am" or "I Am Being Itself." This points to God not as a supernatural person but as the ground of existence itself, pure being, the "is-ness" that underlies all that exists.

This is remarkably similar to how other traditions point to ultimate reality. Brahman in Advaita Vedanta is described as sat-chit-ananda—being-consciousness-bliss. The Tao is the nameless source of being. In each case, what is being pointed to is not a being among beings, but being itself, existence itself, the ground that makes all existence possible.

Ein Sof: The Infinite

In Kabbalah, the ultimate reality is called Ein Sof, which literally means "without end" or "infinite." Ein Sof is completely transcendent, beyond all attributes, beyond all names, beyond all concepts. It is the absolute, undifferentiated unity that precedes all manifestation.

This is identical to the concept of nirguna Brahman (Brahman without qualities) in Advaita Vedanta, or the Tao that cannot be spoken. It is the recognition that ultimate reality transcends all the categories and concepts of the human mind.

But Kabbalah also teaches that Ein Sof manifests through a process of self-limitation and self-revelation. Through a series of emanations called the sefirot (often depicted as the Tree of Life), the infinite contracts itself to allow the finite to exist, the formless takes on form, the hidden becomes revealed.

This is similar to the concept of maya in Hinduism—the creative power of Brahman that allows the one to appear as many. It's similar to the Taoist teaching that being arises from non-being. It's the universal insight that the manifest world is not separate from the divine ground, but is the divine ground expressing itself in form.

The Spark of the Divine

A central teaching in Kabbalah is that every soul contains a spark of the divine, a piece of Ein Sof itself. This neshamah (higher soul) is your true essence, eternally one with God.

This is the same teaching as the Atman-Brahman identity in Advaita Vedanta, the Buddha-nature in Mahayana Buddhism, the "I and the Father are one" in Christian mysticism. Your deepest essence is not separate from the divine. You are not trying to reach God—you are trying to remember that you have never been separate from God.

The Baal Shem Tov, founder of Hasidic Judaism, taught: "In every thing and in every place, the Divine Presence dwells." This is panentheism—the teaching that God is in everything and everything is in God. It's not quite the same as pantheism (God is everything) or theism (God is separate from everything), but points to a non-dual reality in which God and creation are neither identical nor separate.

Tikkun Olam: Repairing the World

Kabbalistic mythology tells of a cosmic catastrophe called the "breaking of the vessels" in which the divine light was scattered and trapped in material forms, creating the world of separation and suffering. The spiritual task is tikkun olam—repairing or healing the world—which involves releasing the trapped sparks of divine light and restoring them to their source.

This can be understood literally as a cosmological myth. But it can also be understood as a metaphor for the spiritual journey. The "breaking" is the fall into separation consciousness, the forgetting of our true nature. The "trapped sparks" are the aspects of divine consciousness identified with limited forms. "Tikkun" is the awakening process—recognizing the divine in all things and seeing through the illusion of separation.

Devekut: Cleaving to God

The highest spiritual attainment in Jewish mysticism is devekut—cleaving to God, or constant communion with the divine. This is not understood as a temporary mystical experience but as a permanent state of consciousness in which one is aware of God's presence in every moment and every action.

This is the same as the Christian teaching of practicing the presence of God, the Islamic teaching of constant dhikr (remembrance), the Buddhist teaching of continuous mindfulness, the Advaitic teaching of sahaja samadhi (the natural state). It is the recognition that separates from God is illusory—God is always present, we are always in God, but we have been distracted by identification with thoughts, emotions, and the separate self.

Rabbi Nachman of Breslov taught: "All the world is a narrow bridge, and the main thing is not to be afraid." The narrow bridge is the razor's edge between form and formlessness, between identification with the separate self and recognition of the divine ground. The teaching is to walk this bridge without fear—to live fully in the world while knowing your true nature transcends the world.

The Misinterpretation in Judaism

Like other traditions, mainstream Judaism has emphasized the exoteric over the esoteric, the law over the spirit, the transcendence of God over the divine immanence.

The Torah, which contains both legal codes and mystical wisdom, has been primarily interpreted through a legalistic lens. Judaism became centered on obedience to 613 commandments, study of rabbinic interpretation, and preservation of Jewish identity and practice.

The mystical teaching that God is the ground of all being and that every soul is a spark of the divine has been marginalized. Kabbalah was traditionally considered too dangerous to study before the age of 40, and was restricted to married Jewish men who had thoroughly studied the legal tradition. The direct path to recognizing one's unity with the divine was hidden behind layers of law, ritual, and exclusivity.

This has resulted in a religion focused on ethnic identity, ritual observance, and intellectual study of texts rather than direct mystical realization. The transcendent God of Sinai who gives commandments has overshadowed the immanent God who is the very ground of existence.

Yet the mystical stream has never disappeared. From the Merkabah mystics of ancient times, through the medieval Kabbalists, to the Hasidic masters, to modern teachers like Rabbi Abraham Joshua Heschel and Reb Zalman Schachter-Shalomi, the living wisdom has been preserved and transmitted.

The Common Core

Examining these diverse traditions—Christianity, Buddhism, Islam, Hinduism, Taoism, and Judaism—reveals a remarkable convergence at the mystical level:

1. Ultimate Reality is One: Whether called God, Brahman, the Tao, Allah, Ein Sof, or Dharmakaya, there is one infinite, eternal ground of all existence.

2. This Reality is Your True Nature: The deepest essence of consciousness is identical with this ultimate reality. The separation between the human and divine is illusory.

3. The Separate Self is an Illusion: What you take yourself to be—the ego, the personality, the isolated individual—is a construction that has no independent existence.

4. Awakening is Possible: Through various practices and ultimately through grace or insight, it is possible to recognize your true nature and live from that recognition.

5. This Recognition Transforms How You Live: Awakening is not an escape from the world but a transformation in how you relate to the world—from separation to unity, from fear to love, from grasping to allowing.

These core insights appear with remarkable consistency across traditions that developed independently in different cultures and time periods. This suggests they are not cultural constructions or religious beliefs, but discoveries about the fundamental nature of consciousness and reality.

The differences between traditions lie primarily in:

- The language and symbols used to point to the truth
- The practices and paths emphasized for reaching realization
- The cultural context and religious framework in which the teaching is embedded
- The degree to which the tradition has been literalized and institutionalized

When we distinguish the essential pointing from the cultural packaging, we find that all authentic mystical traditions are pointing to the same moon, even though they use different fingers.

The Question of Exclusivity

One of the primary obstacles created by religious literalism is the claim to exclusive truth. Christianity has traditionally claimed that salvation is possible only through faith in Jesus Christ. Islam has claimed that Muhammad is the final prophet and the Quran is the ultimate revelation. Judaism has emphasized the special covenant between God and the Jewish people.

These exclusive claims make sense only at the level of the particular religious forms and beliefs. At the level of direct mystical realization, they dissolve. If the truth being pointed to is the non-dual ground of all existence, the infinite awareness that is everyone's true nature, then by definition it cannot be the exclusive property of any group, accessible only through one path, or mediated only by one historical figure.

The enlightened mystics of different traditions have often recognized each other across religious boundaries. There are accounts of Sufi masters and Hindu yogis recognizing each other's realization, of Zen masters and Christian contemplatives finding common ground, of Jewish Kabbalists and Christian mystics sharing insights.

The exclusivist claims arise when the pointers are mistaken for the reality itself, when the map is confused with the territory, when belief in particular propositions is substituted for direct recognition of truth.

This doesn't mean all paths are equally effective or that all teachings are equally clear. Some traditions have preserved the direct pointing more faithfully than others. Some practices are more effective than others for particular individuals. But the truth itself is universal and accessible through multiple paths.

Chapter Seven: The Scientific Worldview and Its Limitations

The Triumph of Science

Modern science represents one of humanity's greatest achievements. Through systematic observation, experimentation, and mathematical modeling, science has unlocked countless mysteries of the natural world. We have split the atom, decoded DNA, landed on the moon, and developed technologies that would seem miraculous to our ancestors.

Science has also liberated humanity from many forms of superstition and dogmatic thinking. It has challenged religious authorities who claimed the earth was the center of the universe, that diseases were divine punishments, that natural disasters were acts of angry gods. It has insisted on evidence, logic, and the willingness to revise beliefs in light of new data.

All of this is to the good. The scientific method is a powerful tool for understanding the behavior of physical systems and developing technologies to manipulate the material world. The problem arises not with science itself, but with scientism—the philosophical assumption that the scientific method is the only valid way of knowing, and that physical matter is the only reality.

The Doctrine of Materialism

Scientific materialism, also called physicalism, holds that:

1. Physical matter is the fundamental reality

2. Consciousness is an epiphenomenon—a byproduct of material processes in the brain

3. The universe is fundamentally mechanical, operating according to physical laws

4. All phenomena can ultimately be explained in terms of particles and forces

5. There is no purpose or meaning inherent in nature—these are human projections

This worldview has become the default assumption of educated people in modern societies. It is taught in universities, assumed in scientific research, and reflected in the materialistic culture that values only what can be measured, quantified, and controlled.

But materialism is not a scientific discovery. It is a philosophical assumption, a metaphysical belief that goes beyond what science can actually demonstrate. And it commits a fundamental error that is strikingly similar to the error of religious literalism: it mistakes the map for the territory, the appearance for the essence.

The Hard Problem of Consciousness

The philosopher David Chalmers has identified what he calls "the hard problem of consciousness." Science can study the neural correlates of consciousness—which brain regions are active during different experiences, how neurons fire, how information is processed. These are the "easy problems" (though they're not actually easy).

The hard problem is: Why is there subjective experience at all? Why does it feel like something to see red, to taste chocolate, to feel sad? How does the objective, third-person world of atoms and molecules give rise to the subjective, first-person world of consciousness?

Materialists assume that consciousness must somehow emerge from complex arrangements of unconscious matter. But this is an assumption, not a demonstrated fact. No one has explained how subjective experience could arise from objective matter, or even what such an explanation would look like.

The alternative possibility—the one pointed to by all mystical traditions—is that consciousness is not produced by matter, but is the fundamental reality. Matter appears in consciousness, not consciousness in matter. The brain does not generate consciousness; rather, consciousness manifests as the brain and everything else.

This is not a return to pre-scientific thinking. It is a recognition that science studies the behavior of appearances within consciousness, but has nothing to say about the nature of consciousness itself. Science operates within the domain of objective experience—the third-person perspective. But consciousness is inherently subjective—the first-person perspective. Using objective methods to study subjective experience is like trying to find the observer by examining the observed.

The Primacy of Experience

There is one thing you know with absolute certainty: you are aware. You have conscious experience. This is the one thing you cannot doubt, because doubting it requires consciousness.

Everything else you think you know—the external world, other minds, the past, the laws of physics—you know only through your conscious experience. You've never actually encountered matter apart from your experience of matter. You've never known a universe that wasn't appearing in your consciousness.

The materialist assumes that consciousness arises within a pre-existing material universe. But you've never actually encountered this pre-existing material universe. All you've ever encountered is your experience, which includes what seems to be a material universe.

This doesn't mean the universe is all in your head, or that nothing exists apart from your personal consciousness. It means that consciousness is primary and matter is secondary—matter is what consciousness appears as, not what consciousness arises from.

This is the insight at the heart of the Yogacara school of Buddhism, which teaches that "mind only" is the fundamental reality. It's the insight of Bishop Berkeley's idealism. It's the implication of quantum mechanics, which reveals that the observer cannot be separated from the observed.

Quantum Mechanics and the Observer

Classical physics operated on the assumption of an objective reality that exists independently of observation. But quantum mechanics has revealed that at the fundamental level, this assumption breaks down.

The famous double-slit experiment demonstrates that particles behave differently depending on whether they are observed. When unobserved, electrons create an interference pattern as if they were waves passing through both slits. When observed, they behave as particles passing through one slit or the other. The act of observation appears to collapse the wave function, determining what had been indeterminate.

Different interpretations of quantum mechanics attempt to make sense of this. The Copenhagen interpretation says that observation collapses the wave function. The many-worlds interpretation says that all possibilities are realized in different branches of reality. The pilot wave theory postulates hidden variables. But all of them point to a reality that is far stranger than the solid, separate objects of classical physics.

The Implications of Modern Physics

Quantum mechanics isn't the only domain of physics that challenges materialist assumptions. Relativity theory revealed that space and time are not absolute containers in which events occur, but are relative to the observer and interconnected as spacetime. Mass and energy are equivalent and interchangeable. The solid, substantial matter of classical physics dissolves into fields, vibrations, and mathematical relationships.

At the quantum level, particles pop in and out of existence. The vacuum is not empty but seething with virtual particles. What seems solid and substantial is actually mostly empty space—atoms are 99.9999999% empty space, with tiny particles somehow held together in patterns by forces we describe mathematically but don't fully understand.

Moreover, quantum entanglement demonstrates that particles can be instantaneously correlated across any distance, suggesting a fundamental interconnectedness that transcends space and separation. Einstein called this "spooky action at a distance," and it has been experimentally verified countless times.

None of this proves that consciousness is fundamental or that mystical traditions are correct. But it does undermine the materialist assumption that we live in a universe of solid, separate objects operating according to mechanical laws. Reality at its deepest levels is far more mysterious, interconnected, and consciousness-involving than classical materialism assumes.

The Limits of the Scientific Method

Science is extraordinarily powerful within its domain—the study of objective, measurable, repeatable phenomena. But this domain has inherent limitations.

Science cannot address questions of meaning, purpose, value, or ethics. It can tell you how things work, but not why they matter or what you should do. It can describe what is, but cannot prescribe what ought to be.

Science cannot fully study consciousness, because consciousness is the subject that does the studying. You can study the objective correlates of consciousness (brain activity, behavior, reports), but you cannot study subjective experience itself from the outside.

Science cannot answer its own foundational questions: Why is there something rather than nothing? Why do the laws of nature take the form they do? Why is the universe comprehensible to the human mind?

Most fundamentally, science operates within the duality of subject and object, observer and observed. It cannot step outside this duality to investigate non-dual awareness itself. This is why the mystical traditions, which use methods of direct introspection rather than external observation, have access to insights that science by its nature cannot reach.

This doesn't mean science and spirituality are incompatible. It means they are complementary—different ways of investigating different aspects of reality. Science explores the objective domain of measurable phenomena. Spirituality explores the subjective domain of consciousness itself.

Reconciling Science and Spirituality

The conflict between science and spirituality arises primarily from two errors:

1. Scientific materialism: The assumption that only physical matter is real and consciousness is merely a byproduct

2. Religious literalism: The belief in supernatural interventions that violate natural law

When we let go of both of these errors, a reconciliation becomes possible.

From the spiritual side, this means:

- Accepting the scientific understanding of natural processes
- Not relying on miracle claims or supernatural intervention
- Recognizing that ancient scriptures contain spiritual wisdom but not scientific knowledge about the material world
- Understanding spiritual teachings as pointers to direct experiential truth, not as competing theories about objective reality

From the scientific side, this means:

- Recognizing the limits of the scientific method
- Not assuming materialism as an unquestioned philosophical framework
- Being open to first-person investigation of consciousness
- Acknowledging that subjective experience is real and primary, even if not fully accessible to objective study

The reconciliation looks like this:

Science is right about the behavior of material phenomena. Evolution, cosmology, neuroscience, physics—these give us accurate knowledge about the patterns and processes of the manifest world.

Spirituality is right about the nature of consciousness and the possibility of awakening from identification with the separate self to recognition of non-dual awareness.

These are not competing claims about the same domain. They are complementary perspectives on different aspects of reality. Science describes the content of experience—the objective world appearing in consciousness. Spirituality points to the context of experience—consciousness itself.

The Participatory Universe

Contemporary physicist John Wheeler proposed that we live in a "participatory universe" in which observers are not separate from what they observe, but are participants in the ongoing creation of reality. This resonates with the mystical insight that consciousness and cosmos are not two separate things, but are aspects of one unified reality.

The mystics go further, suggesting that the entire manifest universe is consciousness itself in form—what appears to be objective matter is actually the self-manifestation of the one awareness. This is not something science can prove or disprove, because it's a recognition that occurs in a different dimension of knowing than scientific investigation.

But science and this mystical vision are compatible. Science can continue investigating the patterns and laws by which consciousness manifests as the material world. Spiritual practice can continue facilitating the direct recognition of consciousness as the ground of all manifestation.

The key is recognizing that they're addressing different questions:

- Science asks: How does the universe behave? What are the patterns and processes?

- Spirituality asks: What is the nature of awareness itself? What am I fundamentally?

Both are valid. Both reveal important truths. Neither has the complete picture alone.

Chapter Eight: The Direct Path to Recognition

Beyond Belief and Practice

We have explored how the enlightened teachings of the world's great traditions have been misinterpreted and literalized into religious dogmas. We have seen how scientific materialism represents another form of spiritual amnesia, forgetting the primacy of consciousness while becoming lost in the content of consciousness.

The question now is: How do we actually recognize our true nature? How do we move from intellectual understanding to direct realization? How do we awaken from the dream of separation?

The first insight is that awakening is not about accumulating beliefs or performing practices. You cannot become what you already are. You can only recognize what has always been true.

All authentic spiritual practices are not about becoming enlightened in the future, but about removing the obstacles to recognizing your true nature in the present. They are subtractive, not additive. They help you see through the illusions that obscure your true nature, not acquire some special status or attainment.

The Direct Path of Self-Inquiry

The most direct method for seeing through these illusions is self-inquiry—the investigation of what you actually are.

This is not an intellectual exercise but a direct looking. You turn attention back on itself to discover the nature of the one who is aware.

Ask yourself: Who am I?

Not as a philosophical question to be answered conceptually, but as a genuine inquiry into the actual nature of your being.

You discover quickly that you are not any object of awareness:

- Not the body (you are aware of it)
- Not the thoughts (you are aware of them)

- Not the emotions (you are aware of them)
- Not the personality (you are aware of it)
- Not the sense of being a person (you are aware of that too)

So what are you? You are the awareness in which all of these appear. You are the witnessing presence, the knowing itself.

But can you find this witness as an object? No—because you are it. The eye cannot see itself. The witness cannot be witnessed. You cannot make awareness an object of awareness, because you are awareness.

This is the final step: recognizing that you are not an object in awareness, but awareness itself. Not a thing that is aware, but awareness in which all things appear.

This awareness is not personal. It doesn't belong to "you." Rather, what you've been calling "you" is a temporary appearance in this awareness, like a wave appearing in the ocean.

The Recognition

What is recognized in awakening? Several things simultaneously:

You have no independent existence: What you've taken yourself to be—the separate self—is seen to be a construction with no independent reality. You are not a thing at all, but the no-thing in which all things appear.

You are not in time: Your true nature has never been born and will never die. The body-mind appears in time, but awareness itself is timeless—always and only now.

You are not in space: You don't exist at a location. Space appears in you; you don't appear in space. Your true nature is the infinite expanse of awareness itself.

You are not separate from anything: The boundary between self and other is seen to be conceptual, not real. All apparent separate things are modifications of the one awareness, like waves on the ocean.

You are what you've been seeking: The peace, fulfillment, and completeness you've been seeking in experiences and attainments is recognized to be your own nature. Not something to be achieved, but what you are.

This recognition is not a new state to be attained. It is the natural state that has always been present but was overlooked. It is like coming home to what you've never left.

The Shift from Seeking to Being

One of the most significant changes that occurs with this recognition is the end of seeking. Not because you've found what you were looking for in the usual sense, but because you recognize that what you've been seeking is what you are.

The seeker is itself part of the illusion. The sense that "I am here, incomplete and lacking, and I need to get there, to enlightenment" is based on the false premise that you are a separate self that needs to be improved or transformed.

When it's recognized that you are already the awareness you've been seeking, the entire project of self-improvement and spiritual attainment is seen through. This doesn't mean you stop growing, learning, or engaging with life. It means the underlying anxiety and sense of lack that drove the seeking is gone.

You no longer need experiences to validate your existence or prove your worth. You no longer need to become someone special or attain some extraordinary state. The ordinary is recognized as extraordinary. The simple fact of being aware, here and now, is recognized as the miracle it has always been.

Common Misconceptions About Awakening

There are several common misunderstandings about what awakening entails:

Misconception 1: Awakening is a permanent state of bliss

Many seekers imagine that enlightenment means being in a constant state of ecstatic joy, forever peaceful and happy. But awakening is not about particular states of experience. States come and go—joy and sorrow, peace and agitation, clarity and confusion. These are all experiences arising in awareness.

Awakening is the recognition of what you are, which is prior to and independent of all changing states. You can be awake and still feel sad. You can be enlightened and still get irritated. The difference is that these states are no longer taken personally. They are recognized as weather patterns in consciousness, not as fundamental to who you are.

Misconception 2: The ego disappears completely

Some people think enlightenment means the total annihilation of the ego or personality. But the functional ego—the capacity to think, plan, remember, and navigate the world—doesn't disappear. What dissolves is the identification with the ego as what you fundamentally are.

The personality continues. You still have preferences, quirks, and characteristics. You still use the word "I" in conventional speech. The difference is that these are recognized as appearances in awareness rather than your identity.

Misconception 3: You gain special powers or abilities

Awakening is not about developing psychic abilities, supernatural powers, or becoming superhuman. These are phenomena that may or may not arise, but they have nothing to do with recognizing your true nature.

In fact, fixation on special powers or unusual experiences is often a subtle form of spiritual materialism—the ego trying to claim something special and maintain its sense of separation.

Misconception 4: Awakening means you never have problems

The recognition of your true nature doesn't exempt you from the challenges of human existence. You still live in a body that ages and can become ill. You still have relationships that can be difficult. You still exist in a world with suffering and injustice.

What changes is your relationship to these challenges. They are no longer resisted as problems that shouldn't be happening. They are recognized as part of the dance of life, to be met with presence rather than avoided or grasped at.

Misconception 5: There's nothing to do after awakening

Some teachings suggest that after awakening, there's nothing left to do—the journey is complete. But awakening is more like a beginning than an end. The recognition of your true nature is just the first step. The integration of this recognition into every aspect of life is an ongoing process.

There is a deepening and stabilization that occurs over time. The initial recognition might be clear but fragile. Through continued inquiry and presence, it becomes more stable and natural. The old patterns of identification don't disappear immediately—they arise with less force and are seen through more quickly.

Pitfalls on the Path

Several common pitfalls can trap seekers and even those who have had genuine glimpses of their true nature:

Spiritual Bypassing: Using spiritual concepts to avoid dealing with psychological issues, emotional wounds, or practical responsibilities. "It's all an illusion" becomes an excuse not to engage with life or address real problems.

Premature Claims of Enlightenment: Having a genuine opening or insight and immediately claiming to be fully enlightened. This often comes from the ego trying to co-opt the recognition and make itself special.

Teacher Inflation: Giving ultimate authority to a teacher and suspending your own discernment. While teachers can be helpful, ultimately you must verify everything through your own direct experience.

Conceptual Understanding Mistaken for Realization: Being able to articulate non-dual philosophy beautifully while still fully identified with the separate self. The map is not the territory.

Nihilism: Concluding that if the separate self is illusory, nothing matters and there are no consequences to actions. This misses the point that while the separate self is illusory, suffering is real, and compassion is the natural expression of recognizing the unity of all beings.

Spiritual Materialism: Collecting experiences, attainments, and credentials as the ego's way of maintaining control. The accumulation of spiritual things—retreats attended, practices mastered, initiations received—becomes another form of seeking and grasping.

Living from Recognition

The real test of awakening is not the profundity of your insights or the power of your experiences, but how you live. Does the recognition of your true nature manifest as greater kindness, authenticity, and presence in daily life?

Presence: When you're not lost in thoughts about past and future, you're naturally present to what is. You hear people more fully. You see beauty more vividly. You respond to situations with fresh awareness rather than conditioned patterns.

Acceptance: When you recognize that what is happening is an expression of the one consciousness, resistance falls away. This doesn't mean passive resignation, but a fundamental acceptance of reality as it is, which actually makes effective action more possible.

Compassion: When the boundary between self and other is seen through, the suffering of others is felt as your own suffering. Compassion arises naturally, not as an obligation or spiritual practice, but as the spontaneous response of recognizing yourself in all beings.

Freedom from Fear: The deepest fears—of death, annihilation, abandonment—are based on the illusion of being a separate, vulnerable self. When that illusion is seen through, these existential fears lose their grip. You still have appropriate caution in dangerous situations, but the underlying anxiety about existence itself dissolves.

Naturalness: You become less concerned with spiritual appearances and more simply yourself. The need to present a spiritual persona, to demonstrate your attainment, or to conform to ideas about how an awakened person should be falls away.

Ordinariness: Paradoxically, awakening makes you more ordinary, not more special. You're less concerned with being impressive and more at ease with being human. The sacred and mundane are recognized as one.

Relationships and Awakening

One of the areas where awakening has the most profound impact is in relationships. When you no longer need others to complete you or validate your existence, relationships become more authentic and free.

You're less prone to projection—seeing in others what is actually unacknowledged in yourself. You're less defensive, because there's less sense of a separate self that needs to be protected. You're more capable of genuine listening, because you're not constantly filtering what others say through the lens of how it affects you.

At the same time, relationships can become more challenging in certain ways. If you've awakened and those around you haven't, there can be a sense of not being understood. The things that used to drive you—achievement, status, accumulation—may no longer motivate you, which can create confusion or conflict with partners, family, or colleagues who still operate from those values.

The key is to meet others where they are, without judgment or the need to convert them to your understanding. You can share your perspective when asked, but there's no compulsion to enlighten everyone. Trust that each person is on their own journey, unfolding in their own time.

Work and Creativity

Awakening doesn't mean you stop working or creating. In fact, many people find that work becomes more effective and enjoyable when it arises from presence rather than from the driven energy of seeking.

The work you do may change, as you're drawn to activities that feel more authentic and aligned with your true nature. Or the same work may continue, but your relationship to it shifts. You're less attached to outcomes, less identified with success or failure, more engaged with the process itself.

Creativity often flourishes after awakening, because you're less inhibited by self-consciousness and concern about how others will judge you. The creative work becomes an expression of the joy of being, rather than a project to prove your worth or gain recognition.

Engagement with the World

A common question is: If everything is recognized as consciousness playing, why bother engaging with social issues, helping others, or trying to reduce suffering?

The recognition of non-duality doesn't negate the relative reality of suffering. While it's true that at the absolute level, there is only consciousness and nothing is ultimately separate, at the relative level, suffering is real and calls for response.

Compassion is not something you do because you should. It's the natural expression of recognizing yourself in all beings. When you see a child suffering, you respond—not because of a belief system that says you should be compassionate, but because the child is you. The boundary between self and other is recognized as illusory, so serving others is serving yourself.

Many awakened beings have been deeply engaged in the world—teaching, serving, creating, advocating for justice. Their engagement comes not from a sense of lack or the need to prove themselves, but from the overflow of presence and love.

The Deepening Process

Awakening is not a one-time event after which everything is permanently different. It's more accurate to say that there is an initial recognition, which then deepens and stabilizes over time.

In the beginning, you might move back and forth between recognizing your true nature and forgetting it, between presence and getting lost in thought. Gradually, the recognition becomes more stable. The old patterns of identification have less grip. The return to presence happens more quickly and easily.

This deepening happens through:

Continued Inquiry: Regularly investigating the nature of awareness and seeing through the subtle ways the sense of separation reasserts itself.

Presence in Daily Life: Bringing awareness to ordinary activities—eating, walking, working, relating—rather than only during formal meditation.

Allowing What Is: Learning to meet whatever arises—thoughts, emotions, sensations, situations—with acceptance rather than resistance.

Integrity: Aligning your actions with your understanding. Living truthfully and authentically.

Gratitude: Recognizing the extraordinary gift of existence itself.

The deepening is not about becoming more enlightened. The recognition permeates more fully into all aspects of life, until there is no division between spiritual and ordinary, between meditation and life.

Chapter Nine: Transcending the Great Dichotomies

Beyond Sacred and Profane

One of the most pervasive illusions maintained by religious literalism is the division between the sacred and the profane, the holy and the mundane, the spiritual and the worldly.

In this view, certain places are sacred (churches, temples, mosques) while others are profane. Certain activities are spiritual (prayer, meditation, religious ritual) while others are worldly. Certain people are holy (priests, monks, saints) while others are ordinary.

But from the perspective of non-dual awareness, everything is equally sacred because everything is an expression of the one consciousness. There is nothing that is not God, nothing that is not Brahman, nothing that is outside the Tao.

This doesn't mean all actions are equally wise or beneficial. It means that the fundamental nature of all that exists is the same divine ground. The cathedral and the kitchen, the meditation cushion and the office chair, the saint and the sinner—all are appearances of the one reality.

When this is recognized, life becomes a seamless whole. You don't have to go to a special place or engage in a special activity to connect with the sacred. The sacred is everywhere, always, because it is the very ground of existence.

This is what the Zen masters meant when they said ordinary mind is the way or chop wood, carry water. Every moment, every action, every breath is an opportunity to recognize and live from the truth of what you are.

Beyond Self and Other

The most fundamental duality is the sense of being a separate self in a world of others. This is the root of all fear, conflict, and suffering.

We spend our lives defending and promoting the interests of me against them. We accumulate possessions, achievements, and experiences to shore up the sense of self. We compete, compare, and judge ourselves against others. We live in constant anxiety about the survival and status of this separate self.

But when it's recognized that the separate self is a construction with no independent existence, this entire drama is seen through. The boundary between self and other is revealed as conceptual, not real.

This doesn't mean you lose your individuality or become merged into some cosmic blob. The unique expression that is your body-mind continues. But it's recognized as one wave among countless waves, all arising from the same ocean.

When the illusion of separation dissolves, several things happen:

Fear diminishes: The existential fear of annihilation loses its foundation, because there was never a separate self to be annihilated.

Love becomes natural: When you recognize yourself in all beings, love is not a feeling you have to cultivate but the natural state of being.

Competition transforms into cooperation: When the boundary between self and other dissolves, serving others is serving yourself.

Judgment softens: You recognize that what you judge in others is often what you haven't accepted in yourself. The other is a mirror.

Beyond Freedom and Determinism

Philosophy has long wrestled with the question of free will: Are we free agents making genuine choices, or are our actions determined by prior causes?

From the perspective of the separate self, this seems like a critical question. If I have free will, I am responsible for my choices. If my actions are determined, then I am not truly responsible.

But from the perspective of non-dual awareness, the question dissolves. There is no separate self that either has or lacks free will. Actions happen, but they are not the actions of an independent entity. They are the spontaneous expression of the totality.

This is the teaching of wu wei in Taoism—action without an actor. It's the teaching of karma yoga in the Bhagavad Gita—action without attachment to the fruits. It's the Zen teaching of acting from no-mind.

When the sense of being a separate doer dissolves, what remains is spontaneous, appropriate action arising from presence. You continue to deliberate, choose, and act, but these are recognized as movements of consciousness rather than the actions of a separate self.

Paradoxically, this brings a deeper sense of freedom—not the freedom of an ego that can do whatever it wants, but the freedom from being bound to the illusion of being a separate self at all.

Beyond Birth and Death

Perhaps the most terrifying duality for the human mind is the division between life and death, existence and non-existence.

We know, intellectually, that the body will die. But the mind recoils from this. We develop elaborate beliefs about afterlives to avoid confronting the possibility of annihilation. We distract ourselves with constant activity to avoid thinking about death.

But when you recognize your true nature as awareness itself, the fear of death loses its grip. The body-mind is born and will die, but awareness was never born and cannot die. It is the eternal background in which birth and death appear.

This is not a belief about immortality of the soul or reincarnation. It's a direct recognition that what you fundamentally are has no beginning or end. You are the timeless now in which the story of birth, life, and death unfolds.

The great Zen master Shunryu Suzuki said, "Each of you is perfect the way you are... and you can use a little improvement." This captures the paradox beautifully. At the absolute level, you are already complete, already perfect, already whole—because you are the infinite awareness itself. At the relative level, there is always growth, development, healing, and transformation.

When both levels are recognized and honored, life becomes a dance between being and becoming, between acceptance and aspiration, between resting in what you are and engaging with what is arising.

The Middle Way

The Buddha taught the middle way—avoiding both extreme asceticism and self-indulgence. But there is a deeper middle way that applies to all the dualities we've explored:

Neither denying the relative reality of the world nor forgetting the absolute reality of consciousness.

Neither escaping into transcendence nor becoming lost in immanence.

Neither rejecting the body and world nor identifying completely with them.

Neither clinging to the form nor collapsing into formlessness.

This middle way is not a compromise or a balance point between two positions. It is the recognition that what appeared to be opposites are actually two sides of one reality. Form is emptiness, emptiness is form. The wave is the ocean, the ocean is the wave.

Living from this recognition means fully engaging with the world while knowing your true nature transcends it. It means caring deeply about the reduction of suffering while recognizing that at the deepest level, nothing is ultimately at risk. It means honoring both the journey and the recognition that you've never left home.

Chapter Ten: Practical Guidance for the Seeker

Starting Where You Are

The spiritual journey begins exactly where you are, with whatever understanding and experience you currently have. There is no prerequisite level of purity, knowledge, or spiritual development required. The recognition of your true nature is available now, regardless of your past or current circumstances.

However, there are practices and approaches that can help create the conditions for this recognition to occur. Think of them not as methods for producing enlightenment, but as ways of removing the obstacles that obscure the truth that is already present.

Meditation: The Foundation

While awakening is not produced by meditation, a regular meditation practice is invaluable for developing the capacity to recognize your true nature.

The essence of meditation is simple: rest as awareness itself. Not doing anything, not trying to achieve any particular state, just being aware of being aware.

A Basic Practice:

Sit comfortably with your spine relatively upright. Close your eyes or lower your gaze.

Notice that you are aware. Not aware of anything in particular—just the simple fact of awareness itself.

Thoughts will arise. Let them. You don't need to stop thinking or achieve a blank mind. Simply notice that you are aware of the thoughts. You are the awareness in which thoughts appear and disappear.

Sensations will arise in the body. Notice that you are aware of these sensations. You are not the sensations—you are the awareness of the sensations.

Emotions may arise. Notice that you are aware of the emotions. The emotions are experiences arising in awareness; they are not what you are.

Rest as this awareness. Not as an effort or concentration, but as a relaxing into what is already present.

When you notice you've been lost in thought (identified with the content of experience rather than resting as awareness), simply return. There is no failure in getting lost, only the opportunity to return again and again.

Practice this for 20-30 minutes daily. Over time, the recognition of yourself as awareness becomes more stable and natural.

Self-Inquiry Throughout the Day

Formal meditation is valuable, but the real practice is bringing this recognition into daily life. Throughout your day, pause and ask yourself: Who is aware right now?

When you're walking, notice: there is awareness of walking. When you're working, notice: there is awareness of working. When you're feeling stressed, notice: there is awareness of stress.

In every experience, you can find the same awareness—not awareness of something, but the awareness itself that is present in all experience.

This simple recognition—I am awareness, not the content appearing in awareness—gradually becomes more natural and continuous.

The Power of Presence

One of the most effective practices is simply being present to what is, without adding commentary, judgment, or storytelling.

When you're drinking tea, just drink tea. Fully present to the warmth, the taste, the sensation. Not thinking about tea, not lost in thoughts about past or future, just present to the actual experience.

When you're listening to someone speak, actually listen. Don't spend half the conversation thinking about what you'll say next, or comparing what they're saying to your own experiences. Just be present to their words, their tone, their presence.

This practice of presence naturally weakens the hold of the thinking mind and reveals the awareness that is always already present.

Working with Difficult Emotions

Spiritual practice is not about escaping difficult emotions or achieving a permanently peaceful state. It's about changing your relationship to all experience, including the challenging stuff.

When a difficult emotion arises—fear, anger, sadness, shame—the typical response is to resist it, distract from it, or identify with it completely. All of these responses strengthen the separate self.

A different approach:

Notice the emotion arising. Name it if helpful: "There is fear" or "There is anger."

Feel it fully in the body. Where is it located? What are the actual sensations? Allow the physical sensations to be fully felt without resistance.

Notice that you are aware of the emotion. The emotion is an experience arising in awareness. It comes and goes, but awareness remains constant.

Rest as the awareness that holds the emotion. You are the space in which the emotion appears, not the emotion itself.

Often, when emotions are fully felt and allowed without resistance or identification, they naturally transform and release. The energy that was locked up in resistance becomes available again.

The Role of Study

While intellectual understanding is not the same as direct realization, study of spiritual teachings can be valuable. Reading or listening to the words of realized beings can point you in the right direction, clarify confusion, and remind you of what you already know but have forgotten.

Useful approaches to study:

Read widely across traditions: The same truth is expressed in many ways. Reading from different traditions helps you recognize the common essence beyond the cultural forms.

Read deeply and slowly: Rather than accumulating many books, spend time with a few key texts. Read them multiple times. Let the words sink in beyond the intellectual level.

Apply what you learn: Don't just collect knowledge. Test the teachings through your own direct investigation. Verify or refute them in your own experience.

Hold teachings lightly: Remember that all teachings are fingers pointing at the moon. Don't mistake the concepts for the reality they point to.

Know when to let go: There comes a point where more reading and study becomes an obstacle. When you find yourself endlessly accumulating spiritual knowledge while avoiding the direct inquiry, it's time to set the books aside and simply be.

Teachers and Teachings

A genuine teacher who has realized their true nature can be immensely helpful. They can point out what you're missing, challenge your assumptions, and reflect back to you the truth of what you are.

However, the teacher-student relationship also carries risks. Here are some guidelines:

Trust your own experience: A true teacher will encourage you to verify everything in your own direct experience, not to simply believe what they say.

Maintain discernment: No teacher is perfect. Even genuinely awakened teachers can have blind spots, personality quirks, or areas where they're not fully integrated. Don't surrender your critical thinking.

Watch for red flags: Teachers who claim to be the only source of truth, who demand unquestioning obedience, who abuse their power sexually or financially, who isolate students from outside influences—these are warning signs of a dysfunctional dynamic.

Recognize projection: You may project onto a teacher qualities they don't actually possess. The peace or clarity you feel in their presence may be a reflection of your own true nature, not something they're giving you. Ultimately, you must recognize this in yourself, not remain dependent on the teacher.

Know when to leave: If the relationship becomes codependent, abusive, or stagnant, trust yourself to move on. The real teacher is your own inner awareness.

Community and Sangha

While the spiritual journey is ultimately individual—no one can recognize your true nature for you—community can provide valuable support.

Benefits of spiritual community:

Mutual support: Others on the path understand the challenges and insights in ways that those who aren't engaged with these questions may not.

Accountability: Regular meetings or practice sessions can help maintain consistency in practice.

Diverse perspectives: Different people will have different insights and experiences that can enrich your own understanding.

Service opportunities: Serving others in the community can be a form of practice and a way of expressing gratitude.

However, be aware of potential pitfalls:

Groupthink: Don't let the community's beliefs replace your own direct knowing.

Spiritual comparison: Resist the temptation to compare your progress to others or create hierarchies of attainment.

Comfort zone: Don't let the community become a spiritual bubble that insulates you from the rest of life.

The ideal is to benefit from community support while maintaining your independence and continuing to engage fully with the wider world.

The Question of Effort

A common confusion on the spiritual path is about the role of effort. Should you make effort to awaken, or does effort itself perpetuate the separate self who is trying to achieve something?

The paradox is that both are true. In the beginning, effort is usually necessary. You need to show up for meditation practice. You need to remember to inquire into your true nature. You need to keep returning to presence when you get lost in thought.

But this effort should be light, like steering a boat rather than rowing furiously. You're not trying to make something happen through force of will. You're creating the conditions and then allowing, trusting, opening.

As the recognition deepens, effort naturally falls away. The practice becomes effortless—not because you're doing nothing, but because what you're doing is aligned with your true nature rather than imposed by the separate self.

The Zen tradition speaks of "effortless effort"—engaged but not striving, committed but not grasping. This is the middle way between laziness and forced striving.

Dealing with Doubt

Doubt is natural on the spiritual path. You may doubt:

- Whether awakening is real or just a concept
- Whether you're capable of awakening
- Whether you're practicing correctly
- Whether your teachers or teachings are authentic
- Whether what you've experienced is genuine or self deception

Doubt can be useful if it leads to deeper inquiry. But it can also become a trap, an endless loop of the mind questioning itself.

The antidote to doubt is direct experience. Don't believe in awakening—investigate it. Look directly at the nature of awareness. Notice what is actually true in your own experience, right now.

Are you aware? Undeniably yes. Can you find a separate self that possesses this awareness? Look closely. What do you actually find?

The truth is not something to be believed. It's something to be recognized, directly and immediately, in your own present experience.

Signs of Progress

While the spiritual path is not about achieving progress in the usual sense, there are indicators that the recognition is deepening:

Increased presence: You spend more time in direct awareness of the present moment and less time lost in thoughts about past and future.

Less reactivity: Situations that would have triggered strong emotional reactions still produce emotions, but they're less intense and don't last as long. You're less identified with the reactions.

Greater ease: There's a sense of relaxation, of not having to maintain or defend a particular self-image. Life becomes less effortful.

Natural compassion: Kindness and compassion arise more spontaneously, without having to cultivate them as deliberate practices.

Decreased seeking: The driving need to achieve, acquire, or become something gradually diminishes. There's more contentment with what is.

Sense of humor: You take yourself less seriously. You can laugh at the whole drama of the separate self trying to become enlightened.

Ordinariness: You feel more human, more ordinary, less concerned with being special or spiritually advanced.

These signs are not goals to pursue, but natural fruits of deepening recognition. If you find yourself trying to cultivate these qualities to prove your spiritual attainment, you've fallen back into the trap of the separate self trying to become something.

When Nothing Seems to Be Happening

Many people on the spiritual path go through periods where it feels like nothing is happening—no insights, no progress, no transformation. This can be discouraging.

Several perspectives on this:

The plateau is part of the path: Progress in any domain doesn't happen in a straight line. There are periods of rapid change and periods of apparent stagnation. Trust the process.

Something may be happening beneath the surface: Like a seed germinating underground before the shoot appears, transformation may be occurring in ways you can't yet see.

Letting go of expectations: The feeling that nothing is happening comes from comparing your present experience to some imagined ideal. What if this, exactly as it is, is what's needed right now?

The trap of spiritual achievement: The separate self wants to achieve awakening, to make progress, to get somewhere. The feeling that nothing is happening may be precisely what's needed to let go of this project.

Return to basics: When you feel stuck, return to simple presence. Put aside all concepts about awakening and just be aware. Notice: you are aware. This is not nothing.

The Importance of Embodiment

Much spiritual teaching emphasizes transcendence—going beyond the body and physical world. But true awakening includes embodiment—recognizing the body and physical life as expressions of consciousness rather than obstacles to overcome.

Practices for embodiment:

Somatic awareness: Regularly bring attention to the felt sense of the body. Notice sensations, movement, breath. The body is happening in awareness; it's not separate from consciousness.

Physical movement: Yoga, tai chi, dance, sports, walking in nature—these can be forms of meditation in motion, ways of being present in the body.

Pleasure and sensuality: Don't reject physical pleasure as somehow less spiritual. Eating, sexuality, the warmth of the sun on your skin—these are consciousness experiencing itself in form.

Grounding practices: When you're too much in the head or lost in abstract spiritual concepts, ground yourself in the physical—feel your feet on the earth, your breath moving, your heart beating.

The goal is not to transcend the body but to recognize the body as an appearance in consciousness, no less sacred than the formless awareness itself.

Integration and Balance

As the recognition of your true nature deepens, it needs to be integrated into all areas of life. This includes:

Relationships: Bringing presence and authenticity to how you relate to others.

Work: Engaging with your vocation from presence rather than compulsion.

Health: Taking care of the body as the vehicle for consciousness in this life.

Finances: Handling practical matters responsibly while not being driven by fear or greed.

Service: Finding ways to contribute to the "reduction" of suffering in the world.

Rest and play: Allowing time for rest, joy, and activities that have no purpose beyond themselves.

True awakening is not about escaping into some transcendent realm. It's about being fully present to life exactly as it is, while knowing your true nature transcends all of it.

The paradox is complete: You are not the body-mind, yet you are embodied. You are not in the world, yet you are fully engaged with the world. You have never been born and will never die, yet you live this precious, temporary human life.

Living from this paradox is the art of awakened existence.

Chapter Eleven: The Synthesis - Reconciling All Perspectives

The Unity Behind Diversity

We have journeyed through the great spiritual traditions of humanity—Christianity, Buddhism, Islam, Hinduism, Taoism, and Judaism—and seen how each, at its mystical core, points to the same essential truth: the non-dual nature of reality and the possibility of awakening from the illusion of separation.

We have also examined how these teachings have been systematically misinterpreted and literalized, creating religious institutions that often obscure rather than reveal the truth they were meant to transmit.

And we have explored how modern scientific materialism, despite its tremendous practical successes, commits a similar error—mistaking the appearance for the essence, the content of consciousness for consciousness itself.

What emerges from this exploration is a vision of reality that honors both the diversity of perspectives and the underlying unity:

At the absolute level: There is only consciousness—infinite, eternal, undivided awareness expressing itself as everything that exists. All apparent separate things are modifications of this one reality, like waves on the ocean or dreams in the mind of the dreamer.

At the relative level: The manifest universe operates according to discoverable patterns and laws. Science's descriptions of these patterns are valid within their domain. The diversity of spiritual traditions reflects different cultural expressions of the same essential insight.

At the practical level: We live as human beings in a world where suffering is real and calls for compassionate response. The recognition of non-duality doesn't negate the relative reality of the world; it transforms our relationship to it.

Resolving Apparent Contradictions

Many of the apparent contradictions between spiritual traditions, and between spirituality and science, dissolve when we recognize that they are addressing different levels or aspects of reality.

Is God personal or impersonal? Both. The ultimate reality is beyond all attributes (impersonal), yet it manifests as all forms, including forms that can be related to personally. Devotional practice to a personal deity and recognition of the impersonal absolute are not contradictory—they are different approaches appropriate for different temperaments and levels of understanding.

Is there free will or determinism? Neither and both. There is no separate self that possesses free will, yet actions and choices happen. The question itself assumes a separate self that either has or lacks free will. When that assumption is seen through, the question dissolves.

Is the world real or illusory? Both. The world is real as an appearance in consciousness. It is illusory insofar as it appears to be composed of separate, independent objects existing apart from consciousness. The rope is real; the snake you mistake it for is illusory.

Should we engage with the world or renounce it? Both. Engage fully with life while knowing your true nature transcends all forms. Serve others while recognizing there are ultimately no others. Care deeply about reducing suffering while knowing that at the deepest level, nothing can be damaged.

These apparent contradictions arise from trying to capture non-dual truth in dualistic language. When you recognize the truth directly, the contradictions resolve naturally.

The Three Levels of Teaching

Different spiritual teachings address different levels of understanding. Recognizing this helps reconcile apparent contradictions:

Level 1 - Dualistic Practice: For those operating from complete identification with the separate self, teachings emphasize moral behavior, devotional practice, accumulating merit, and gradual purification. This is the level of conventional religion.

These teachings are not wrong—they are appropriate for this level. They help reduce harmful behavior, cultivate beneficial qualities, and gradually prepare the ground for deeper understanding.

Level 2 - Progressive Path: For those beginning to question the separate self, teachings emphasize meditation, inquiry, and practices designed to weaken identification with thoughts, emotions, and the body-mind. This is the level of most serious spiritual practice.

Again, these teachings are valid. They help develop the capacity to recognize what you are by removing the obstacles.

Level 3 - Direct Pointing: For those ready to hear it, teachings point directly to what you already are. No practice is needed, no gradual path required. Simply recognize: you are awareness itself, here and now. This is the level of the most radical non-dual teachings.

The confusion arises when teachings from different levels are mixed up or when someone at one level tries to apply teachings meant for another level. A person deeply identified with the separate self who tries to apply the highest teachings ("there is no one to practice, nothing to do") may simply reinforce spiritual bypassing and laziness. A person ready for direct recognition who gets caught up in endless practices and moral purification may miss the immediacy of what's already present.

The skillful approach is to meet yourself where you are while remaining open to what's possible.

Science and Spirituality as Complementary

We can now see how science and spirituality, properly understood, are not in conflict but are complementary ways of knowing:

Science investigates: The patterns and processes of the manifest world, the laws that govern the behavior of forms, the mechanisms by which consciousness appears as the material universe.

Spirituality investigates: The nature of consciousness itself, the ground of all experience, the subject that cannot be made into an object of study.

Science asks: How does the universe behave? What are the patterns?

Spirituality asks: What is the nature of the awareness that knows the universe?

Science operates in the third-person perspective—objective investigation of phenomena.

Spirituality operates in the first-person perspective—direct investigation of the subject of experience.

Both are valid. Both reveal important truths. Science without spirituality leads to materialism and the forgetting of consciousness. Spirituality without science can lead to superstition and rejection of the natural world. Together, they provide a complete picture.

The Evolution of Consciousness

One way to understand the human journey is as an evolution of consciousness through different stages:

Unconscious Unity: Before self-awareness develops, there is a kind of unconscious unity with nature, with the present moment. This is the state of infants and animals—no sense of separation, but also no self-awareness.

Separation and Self-Consciousness: As the ego develops, the sense of being a separate self crystallizes. This is necessary for functioning in the world, but it comes with the pain of feeling isolated, vulnerable, and mortal. This is where most of humanity operates.

The Search: At some point, the pain of separation becomes conscious. We begin searching for something to fill the void—through relationships, achievements, beliefs, experiences. This search is ultimately the search for our true nature, though we don't recognize it as such.

Glimpses and Insights: Through various means—spiritual practice, suffering, grace, psychedelics, or spontaneous experiences—we catch glimpses of something beyond the separate self. These glimpses are often temporary and can be confusing, but they reveal that another way of being is possible.

Recognition and Awakening: There is a shift from identification with the separate self to recognition of awareness itself as what we are. The illusion of separation is seen through. This is not a return to unconscious unity but conscious unity—recognizing our identity with the source while retaining the capacity for self-reflection.

Integration and Embodiment: The recognition deepens and becomes integrated into all aspects of life. We learn to live as awareness while fully engaged with the human experience. This is an ongoing process without end.

This is not a linear progression that everyone follows in the same way. There are leaps forward and apparent steps backward. Different people may spend more or less time at different stages. But understanding this general arc can help make sense of the journey.

The Role of Suffering

Suffering plays a paradoxical role in the spiritual journey. On one hand, the goal is to end suffering through the recognition of our true nature. On the other hand, suffering is often what drives us to seek truth in the first place.

As long as the separate self seems basically functional and life is relatively comfortable, there may be little motivation to question the fundamental assumptions about identity and reality. But when suffering becomes acute—through loss, illness, existential crisis, or simply the accumulated weight of living with the illusion of separation—we are forced to look deeper.

This is why many spiritual traditions speak of the gift of suffering. Not because suffering is inherently good, but because it can break through the complacency that keeps us locked in illusion.

The Buddha's first noble truth—that there is suffering inherent in conditioned existence—is not pessimism. It is clear-seeing. As long as we are identified with the separate self, there will be suffering. This suffering is not a punishment or a problem to solve through external means. It is a signal pointing us toward the recognition of our true nature.

In this sense, suffering can be a fierce grace—the universe's way of waking us up from the dream of separation.

The Question of Purpose

If the separate self is illusory and everything is already perfect as an expression of the one consciousness, what is the purpose of this whole drama? Why does consciousness create the appearance of separation, suffering, and the journey back to itself?

Many answers have been offered:

Lila: The Hindu concept of divine play—consciousness creates the universe for the sheer joy of it, the way an artist creates or a child plays, without needing any reason beyond the creative expression itself.

Evolution: Consciousness is evolving, exploring itself through infinite forms and perspectives, developing ever-greater complexity and self-awareness.

Love: The Sufi view that consciousness creates apparent separation so that it can experience the joy of reunion, the lover discovering the Beloved.

No reason: The Zen perspective that asking why is itself a trap of the conceptual mind. There is no meaning outside the meaning we create. Reality just is.

All of these perspectives may be valid, or none of them. The question of ultimate purpose may be one that the human mind cannot answer, because purpose itself is a concept that arises within consciousness, not something that can be attributed to consciousness itself.

What we can say is: Here we are. Conscious, alive, aware. Whatever the ultimate reason, the immediate response is to recognize what we are and to live from that recognition with as much presence, love, and wisdom as possible.

The End of Seeking

One of the most profound shifts that occurs with awakening is the end of seeking. Not because you've found what you were looking for in the conventional sense, but because the seeker itself is seen through.

The separate self is the seeker. It seeks experiences, achievements, recognition, and even enlightenment to shore up its sense of existence and prove its worth. As long as the separate self is running the show, there is always something more to seek, some lack to fill.

But when the separate self is recognized as a construction with no independent existence, seeking falls away naturally. Not as a discipline or a decision, but as a simple recognition that what you've been seeking is what you are.

This doesn't mean you stop having preferences or pursuing goals. It means the underlying sense of lack and the compulsive need to fill it dissolves. You can engage with life fully while not needing anything from it to validate your existence.

This is the peace that "passes understanding"—not a peaceful state to be achieved, but the recognition of the peace that has always been present, overlooked in the search for it.

Living the Truth

Ultimately, the teaching must be lived, not just understood intellectually. The recognition of your true nature must permeate every aspect of life—relationships, work, challenges, joys, sorrows.

This is not about becoming perfect or transcending your humanity. It's about being fully human while knowing that your true nature transcends the human.

It's about making mistakes and learning from them, experiencing sadness and allowing it, feeling joy and appreciating it—all while knowing that these are movements in consciousness, not defining features of who you are.

It's about serving others not from moral obligation but from the natural overflow of recognizing yourself in all beings.

It's about engaging fully with life while holding it lightly, caring deeply while not clinging, being committed while remaining free.

This is the art of awakened living—the integration of the absolute and relative, the transcendent and immanent, being and becoming.

Conclusion: The Invitation

Beyond All Words

We have traveled together through the great religious traditions, through the halls of science and philosophy, through the landscape of spiritual practice and awakening. We have examined how the living truth has been systematically misinterpreted and literalized, creating the twin illusions of religious dogmatism and scientific materialism.

But ultimately, all of these words are pointing to something that cannot be captured in words. The truth we are discussing is not a philosophical position to be defended, not a belief system to be adopted, not even an experience to be attained.

It is what you are, right now, reading these words. It is the awareness in which these words appear and are understood. It is the consciousness that has been present in every moment of your life, unchanged by all the changes in experience.

This awareness is not personal—it doesn't belong to you. Rather, what you call "you" is a temporary appearance in this awareness, like a wave on the ocean, a character in a dream.

The Direct Recognition

Right now, in this very moment, you can recognize this truth directly.

Notice that you are aware. Not aware of anything in particular—just the simple, undeniable fact of awareness itself.

This awareness is not something you do. You don't create it or maintain it through effort. It is simply what is present, what has always been present, what cannot not be present.

All experience arises in this awareness—thoughts, sensations, perceptions, emotions. They come and go, but awareness remains constant.

Look for the one who is aware. Can you find a separate self, a subject that possesses this awareness? Or is there simply awareness itself, in which the sense of being a separate self arises as one more experience?

If you look honestly and carefully, you will not find a separate self. You will find only awareness—open, empty, spacious, timeless.

This awareness is what you are. Not what you will become through practice or attainment, but what you have always already been, overlooked in the search for it.

This is the great secret hidden in plain sight: You are That which you have been seeking. The kingdom of God is within you. Your true nature is Buddha-nature. You and the Father are one. Atman is Brahman. You are the Tao.

These are not metaphysical claims to be believed. They are invitations to look directly and see for yourself.

What Changes and What Doesn't

When this recognition occurs, what changes?

Your circumstances may or may not change. Your personality may remain largely the same. Your challenges and responsibilities continue. The body still ages and will die.

What changes is the center of gravity, the fundamental sense of identity. Instead of being a separate self trying to navigate the world and find peace and fulfillment, you recognize yourself as the awareness in which the entire drama arises.

This shift is subtle and profound. Life continues, but from a different ground. Actions happen, but there's no separate doer.

Experience arises, but there's no separate experiencer. Thoughts come and go, but there's no separate thinker that they belong to.

There is just this—consciousness appearing as everything, aware of itself through infinite perspectives, expressing itself in infinite forms.

And somehow, miraculously, one of those forms is reading these words right now, recognizing itself in them.

The Pathless Path

There is no path to what you are, because you cannot travel to where you already are. All the practices, teachings, and traditions are not creating something new but removing the obstacles to recognizing what has always been present.

In this sense, the spiritual path is pathless. You are not going anywhere, becoming anything, or attaining any state. You are simply recognizing the truth of what you are, here and now.

Yet paradoxically, practices and teachings can be valuable. They help shift attention from the content of consciousness to consciousness itself. They weaken the habitual identification with thoughts and emotions. They create space for the recognition to occur.

Use them, but don't depend on them. Practice, but don't make practice into a project of the separate self trying to achieve enlightenment. Study, but don't mistake concepts for reality. Seek teachers, but ultimately trust your own direct knowing.

And when the recognition occurs—whether gradually or suddenly, whether dramatically or quietly—allow it. Don't try to maintain it or make it into something special. Don't claim it as an achievement or use it to feel superior to others.

Simply rest as what you are. Live from this recognition. Allow it to permeate every aspect of your life.

The End and the Beginning

This book ends, but your journey continues. Or perhaps more accurately, the journey is recognized as never having begun—because you have never actually been separate from the source, never actually lost, never actually needed to be saved.

The entire drama of seeking and finding, of losing yourself and waking up, has been consciousness playing with itself, the divine dancing with its own reflection.

You are free to continue the dance, but now you know what you are. The character remains, but you know yourself as the author. The wave continues, but you know yourself as the ocean. The dream goes on, but you know yourself as the dreamer.

This is the great liberation—not liberation from life, but liberation in and as life itself.

The Final Invitation

If you have read this far, something in you recognizes the truth being pointed to. Whether you call it curiosity, resonance, or the pull of your true nature, you are being called to wake up from the dream of separation.

My invitation to you is simple:

Look directly at the nature of your own awareness. Not at the contents of awareness—the thoughts, sensations, perceptions—but at awareness itself.

Notice that you are aware. Rest as that awareness. Return to it again and again until the return becomes continuous, until the recognition is stable, until there is no longer any separation between the awareness and what appears in it.

This is not a doing but an allowing. Not an achievement but a recognition. Not becoming something new but realizing what has always been true.

The truth that sets you free is not hidden in ancient scriptures or secret teachings. It is the most obvious fact of your existence: you are aware. You are the awareness in which all experience arises.

This awareness is not separate from the divine ground of all existence. It is the divine ground, appearing as you, aware of itself through your eyes.

See this. Know this. Be this.

And live from this recognition with compassion, presence, and love.

This is the teaching. This is the practice. This is the realization.

May you recognize what you are.

May you awaken from the dream of separation.

May you know the peace that has always been present.

May you live as the love that you have always been.

The journey is complete. You are home. You have never left.

Welcome.

Epilogue: Practical Exercises and Daily Reminders

Exercise 1: The Simple Recognition

Notice that this awareness is not something you do or create. It is simply what is present.

Rest as this awareness for a few moments.

Return to your activity, but with the recognition that you are the awareness in which the activity appears, not just the character performing the activity.

Do this practice dozens of times throughout the day. Each time, you're strengthening the recognition of what you truly are.

Exercise 2: Investigating the Separate Self

Set aside 20 minutes in a quiet place.

Ask yourself: Where is the "I" that I take myself to be?

Look for it systematically:

Is it in the body? Notice that you are aware of the body. You can observe bodily sensations. So the body is an object of awareness, not the subject.

Is it in the thoughts? Notice that you are aware of thoughts. They come and go. You observe them. So thoughts cannot be what you are.

Is it in emotions? Notice that you are aware of emotions arising and passing. They are observed, not the observer.

Is it in the sense of being a continuous person over time? Notice that this continuity is constructed from memories (which are present thoughts about the past) and projections (which are present thoughts about the future). The actual experience is always now.

Where, then, is this separate self? Can you actually find it?

What remains when you can't find a separate self? Only awareness itself—the witnessing presence that is prior to all experiences.

Rest as this awareness for the remainder of the practice period.

Exercise 3: Presencing in Activity

Choose a routine daily activity—washing dishes, taking a shower, eating a meal, walking.

During this activity, be completely present to the direct sensory experience.

If washing dishes: Feel the temperature of the water. Notice the texture of the soap, the plates, the sponge. Hear the sound of water running. See the play of light on the bubbles.

Notice when thoughts arise that take you away from the direct experience—thoughts about the past or future, judgments, commentary. Gently return to direct sensory presence.

Notice that you are the awareness in which these sensations appear. You are not in the experience; the experience is in you.

Practice this with different activities throughout the day. Each activity becomes an opportunity to recognize yourself as awareness rather than getting lost in the story of being a separate person doing things.

Exercise 4: Meeting Difficult Emotions

When a difficult emotion arises—fear, anger, sadness, shame:

Don't try to escape it or fix it. Don't suppress it or indulge in the story about it.

Instead, turn toward it with interest and curiosity.

Feel it fully in the body. Where is it located? What are the actual sensations? Heat? Tightness? Heaviness? Tingling?

Notice that you are aware of these sensations. The emotion is an experience arising in awareness, not what you are.

Notice any thoughts associated with the emotion. Recognize them as thoughts, not truths.

Rest as the awareness that holds both the sensations and the thoughts. You are the space in which the emotion appears, not the emotion itself.

Often, when fully felt and allowed without resistance or identification, emotions naturally transform and release their energy.

Exercise 5: Seeing Through the Story

Several times a day, notice the stories you're telling yourself:

About who you are: "I'm someone who is good at this but bad at that, who succeeded here but failed there."

About others: "She thinks this about me. He always does that. They don't understand."

About situations: "This shouldn't be happening. Things should be different. If only..."

Recognize these as stories—mental constructions, not reality itself.

Ask: What is actually here, prior to the story? What is my direct experience before I interpret it?

Notice that the stories create suffering. The direct experience, without the story, is workable.

Notice who is aware of the stories. This awareness is not caught up in the stories. It is free, spacious, at peace.

Rest as this awareness rather than believing the stories.

Daily Practice Schedule

A suggested rhythm for integrating these practices:

Morning (20-30 minutes):

- Sit in meditation, resting as awareness
- Practice self-inquiry: "Who am I?"
- Set an intention to remember your true nature throughout the day

Throughout the Day:

- Practice "The Simple Recognition" frequently—every hour if possible
- Bring full presence to at least one routine activity
- When difficult emotions arise, use them as opportunities to practice

Evening (15-20 minutes):

- Review the day: When did you remember your true nature? When did you forget?
- Release any judgments about how well you practiced
- Rest in gratitude for the gift of awareness itself

Before Sleep:

- As you're falling asleep, rest as awareness
- Notice that even as thoughts slow down and sleep approaches, awareness remains

This is not a rigid schedule but a framework. Adapt it to your life circumstances. The key is consistency—daily practice, even if brief, is more valuable than occasional marathon sessions.

Common Questions and Responses

Q: How long does it take to awaken?

A: This question assumes that awakening is something that happens in time to a person. But awakening is the recognition that what you are is timeless. It can happen in an instant, or the recognition can gradually stabilize over years. The seeking mind wants a timeline, but that very seeking is part of what you're waking up from. The only moment that matters is now.

Q: How do I know if I'm making progress?

A: Progress is tricky to measure because the goal is not to become something different but to recognize what you already are. That said, signs include: more frequent presence, less reactivity, natural compassion, decreased seeking, and a sense of ease. But don't make these into goals to achieve or you'll be back in the trap of the separate self trying to become enlightened.

Q: I had a powerful awakening experience, but it faded. Did I lose it?

A: You can't lose your true nature—it's what you are. What happened is that the recognition faded and identification with thoughts and the separate self reasserted itself. This is normal. The recognition typically fluctuates before it stabilizes. Each time you recognize and return, the recognition becomes more stable.

Q: I understand this intellectually, but I don't feel awakened. What am I missing?

A: The shift from intellectual understanding to direct recognition is subtle but crucial. You're missing the felt sense of being awareness rather than just thinking about awareness. This is why practices that bypass the conceptual mind—like self-inquiry, meditation, or simply pausing to notice that you're aware—are valuable. The recognition is not a thought but a shift in the center of gravity.

Q: Should I quit my job, leave my relationship, or make major life changes after awakening?

A: Awakening itself doesn't require any particular lifestyle. Some people's external circumstances change naturally as they align with deeper authenticity. Others continue with the same job and relationships but relate to them differently.

Don't make major changes from a concept of how an awakened person should live. Let any changes arise organically from clear seeing, not from ideas about spirituality.

Q: What about psychedelics—can they facilitate awakening?

A: Psychedelics can provide glimpses of non-dual awareness and loosen the grip of the separate self. For some people, these experiences are valuable catalysts. However, glimpses are not the same as stable recognition. And psychedelics can also create spiritual materialism—collecting experiences while the separate self remains intact. They are tools, not shortcuts. The recognition must ultimately stabilize without chemical support.

Q: I feel like I'm going crazy. Is this normal?

A: As the sense of being a solid, continuous self begins to dissolve, it can feel disorienting or even frightening. The structures that have defined your identity are being seen through. This is sometimes called the "dark night of the soul." It's a phase in the journey. If you're experiencing severe distress, seek support from a teacher, therapist, or trusted guide. The dissolution of the false self can be turbulent, but what's on the other side is freedom.

Q: Is awakening the same in all traditions, or are there different types?

A: The fundamental recognition—that awareness is your true nature and the separate self is illusory—is the same across traditions. However, different traditions emphasize different aspects and use different language. Some emphasize emptiness, others fullness. Some emphasize transcendence, others immanence. These are different facets of the same jewel, not different jewels.

*lQ: What about karma and reincarnation? Are they real?

A: This depends on what level you're asking from. At the absolute level, there is no separate soul to be reborn. At the relative level, the teaching of karma points to the truth that actions have consequences. Whether these consequences extend beyond one lifetime is not something that can be verified through direct experience. The important question is: Who is it that would be reborn? When you can't find a separate self, the question dissolves.

Q: After awakening, do you still have preferences and desires?

A: Yes. The body-mind continues to have preferences—it prefers comfort to pain, health to illness, beauty to ugliness. Desires still arise. What changes is the relationship to preferences and desires. They're held lightly rather than grasped at desperately. You can prefer one thing while accepting what is. You can desire something while being complete whether you get it or not.

Q: How do I help others awaken?

A: You can point others toward truth through your words and through your presence. But you cannot awaken anyone else—they must recognize their own true nature. The most powerful teaching is living from your own recognition. When you embody presence, peace, and love, others naturally sense it and may be drawn to investigate. Share when asked, but don't try to convert or convince. Trust that each person's awakening unfolds in its own time.

Warning Signs: When Practice Goes Off Track

Be aware of these red flags that indicate your practice may have gone off course:

Spiritual Bypassing: Using spiritual concepts to avoid dealing with psychological issues, trauma, or practical responsibilities. "It's all an illusion" should not become an excuse for neglecting real problems.

Inflation: Feeling superior to others because of your spiritual understanding or experiences. This is the ego co-opting spirituality to feel special.

Dissociation: Becoming so detached from emotions and the body that you're essentially numb or disconnected. True awakening includes all of experience, not rejection of it.

Perfectionism: Expecting yourself to be constantly peaceful, never reactive, always present. This creates a new form of self-judgment and seeking.

Isolation: Withdrawing from relationships and normal life because you feel others don't understand. While some solitude is valuable, extreme isolation can be a form of spiritual bypassing.

Dependency: Becoming overly reliant on a teacher, teaching, or practice to maintain your sense of well-being. True awakening is recognizing what you are independent of any external support.

If you notice these patterns, bring them into awareness with honesty and compassion. Consider working with a therapist or counselor in addition to spiritual practice. The goal is wholeness, not transcendence at the expense of being human.

Resources for Continued Exploration

While the truth is found in direct recognition rather than books, certain texts and teachers can be valuable pointers. Here are some suggestions:

Classic Texts:

The Upanishads - The ancient Indian texts that contain some of the clearest non-dual teachings ever written.

The Bhagavad Gita - Combines devotional and non-dual teachings in a powerful synthesis.

The Tao Te Ching by Lao Tzu - Simple, profound pointers to the nameless source.

The Heart Sutra - Brief Buddhist text on emptiness and form.

The Gospel of Thomas - Early Christian text with sayings of Jesus that preserve more of the non-dual essence.

The Cloud of Unknowing - Medieval Christian mystical text.

The Works of Meister Eckhart - Radical Christian mysticism.

The Poetry of Rumi - Sufi teachings expressed through ecstatic poetry.

Modern Teachers and Texts:

"I Am That' by Nisargadatta Maharaj - Direct non-dual teachings from a modern sage.

'Be As You Are' - The teachings of Ramana Maharshi on self-inquiry.

'The Power of Now* by Eckhart Tolle - Accessible introduction to presence.

*'The End of Your World' by Adyashanti - Guidance on the awakening process.

'The Way of Liberation' by Adyashanti - Practical guidebook for the direct path.

'Emptiness Dancing' by Adyashanti - Teachings on the nature of awakening.

'The Headless Way' by Douglas Harding - Experimental approach to recognizing no-self.

'The Book of Not Knowing' by Peter Ralston - Exploring the nature of consciousness.

'Being Aware of Being Aware' by Rupert Spira - The direct path of awareness.

Different Approaches:

Different teachers and traditions emphasize different methods. Explore to find what resonates:

Self-Inquiry (Ramana Maharshi, Nisargadatta Maharaj): Investigating "Who am I?" until the separate self is seen through.

The Direct Path (Atmananda Krishna Menon, Rupert Spira, Francis Lucille): Recognizing awareness as your true nature through direct investigation.

Zen (Many teachers): Using meditation, koans, and direct pointing to break through conceptual mind.

Dzogchen/Mahamudra (Tibetan Buddhism): Recognizing the nature of mind directly.

Neo-Advaita (Tony Parsons, Nathan Gill): Radical teaching that there is no one to practice and nothing to do.

Contemporary Embodied Approaches (Judith Blackstone, Reggie Ray): Integrating awakening with somatic awareness and trauma healing.

Final Words

This book has been a finger pointing at the moon. The moon itself—the direct recognition of your true nature—cannot be given through words. It can only be recognized by you, in your own immediate experience.

Everything written here is provisional—useful pointers that should ultimately be set aside. Don't collect these teachings as more spiritual knowledge. Use them as tools for investigation, then let them go.

The invitation remains: Look directly at what you are. Not what you think you are, not what you've been told you are, but what you directly experience yourself to be.

Are you aware? Obviously yes.

Can you find a separate entity that possesses this awareness? Look carefully.

What remains when no separate self can be found? Only awareness itself—open, spacious, timeless, free.

This is what you are. This is what has been looking through your eyes your entire life. This is what will remain when the body dies. This is the eternal within the temporal, the infinite within the finite, the divine within the human.

You are That.

Not as a belief or concept, but as the most immediate, obvious, undeniable fact of your existence.

See this. Know this. Be this.

And live from this recognition with presence, compassion, and love.

The rest unfolds naturally.

A Closing Blessing

May these words serve their purpose and then be forgotten.

May you recognize what you are beyond all words and concepts.

May you know the peace that has always been present.

May you live as the love that you have always been.

May all beings awaken from the dream of separation.

May all suffering be transformed in the light of recognition.

May the truth that liberates be realized by all.

Om shanti shanti shanti
Peace peace peace
The teaching is complete.
The recognition is immediate.
You are home.

About This Book

This book emerged from the recognition that the core truth taught by all genuine spiritual traditions is the same: the non-dual nature of reality and the possibility of awakening from the illusion of separation. Yet this truth has been systematically obscured by religious literalism and scientific materialism—two different forms of forgetting our true nature.

By examining how the enlightened teachings of Christianity, Buddhism, Islam, Hinduism, Taoism, and Judaism have been misinterpreted, and by showing how science and spirituality can be reconciled rather than opposed, this book aims to point directly to the living truth that precedes all traditions and conceptual frameworks.

The hope is not that you believe what is written here, but that you investigate for yourself and recognize directly what you are.

If this book serves that purpose for even one reader, it has fulfilled its function.

May you awaken to what you are.

May you live from that recognition.

May you be free.

About the Author

Mognu knows his role to be he who bridges the gap between reality and unreality, truth and untruth, matter and spirit, for those who are seeking it. Mognu also fills the vast space of the misunderstanding of the truth of gnosis and its illumined revelators throughout history. All writings by Mognu are a product of direct experience. Direct knowledge. Be it fiction or nonfiction, all writings are based upon Mognu's personal experience. The purpose of Mognu's writings is to teach about the intricacies and history of Gnosis.

www.ingramcontent.com/pod-product-compliance
Lightning Source LLC
LaVergne TN
LVHW022303200926
848481LV00004B/1204
* 9 7 9 8 2 3 3 7 7 9 2 7 5 *